UNEXPECTED PRISONER

MEMOIR OF A VIETNAM POW

ROBERT WIDEMAN
AND CARA LOPEZ LEE

Praise for
UNEXPECTED PRISONER

"Unless you were there, you will never truly understand what it was like to be fighting a war that had no description. As a Vietnam Veteran I can empathize with all who have served, but reading Robert's story has added a whole new dimension to this experience. I had to engage the enemy on occasion, but I never had to live with him. I can only imagine what it must have been like to be a POW in North Vietnam. Robert Wideman and his remarkable book do an incredible job of putting us there!"

-**W.R. (Bill) Cobb**, 1st Army Vietnam Veteran, Fortune 50 Executive, and Author

"I roomed with Bob at various times during those trying and difficult years as POWs in North Vietnam. I shared a number of the episodes he relates. His perceptive analyses and detailed descriptions vividly depict our interactions and emotions as human beings unwillingly thrown together into inescapable, trying situations and our individual and collective efforts to cope with and overcome these adverse conditions."

-**Howard Hill**, Retired U.S. Air Force Colonel

"An eye-opener. I had developed a totally different picture of POW existence. Unexpected Prisoner is a must-read."

-**Billy Thornton**, PhD, Vietnam War Veteran

"This is a truly remarkable account of experiences from within the walls of captivity."

–**Rick Fischer**, Vietnam War Veteran, Army Pathfinder shot down in 1969

"A lot of great books have been written about the American experience that is called Vietnam. Robert Wideman's Unexpected Prisoner is a great book that tells the truth regardless of the consequences to a man who was shot down and held captive for six long years."

-**Albert St. Clair**, Navy Diver, Carrier Pilot and Veteran of Iraqi Freedom, Desert Shield, and Enduring Freedom

"Unless you were there, you will never truly understand what it was like being a POW in Vietnam. Robert Wideman and his remarkable book do an incredible job of putting you there!"

-**Brad Hoopes**, Author of *Reflections of Our Gentle Warriors*

"Robert Wideman and I share a common bond from our years in Vietnam. He was shot down and taken captive. I was overrun at Con Thien (a hill at the DMZ). His story of captivity is the most accurate version of the events that occurred in the North I have ever heard. Truly refreshing."

-**Captain William (Willydog) Roberts**, Retired U.S. Marine

"As a naval aviator who endured a very real ten-day survival training exercise, I can barely imagine six years as a POW! Thanks to Robert for sharing. This book is a must-read for those interested in understanding the risks our men and women in uniform with combat assignments and enemy exposure face every single day."

-**Sam Solt**, Former Naval Aviator

"As a member of U.S. Army Counterintelligence, part of my job was acting as the 'Aggressor' for escape and evasion training of Army personnel. Robert Wideman's book lays out in spades what we tried to prepare our soldiers to expect and endure if captured. This book should be a text for training in the future."

-**Howard Lowell**, Former Army Counterintelligence Officer

Robert Wideman/Unexpected Prisoner
Printed in the United States of America

Unexpected Prisoner/ Robert Wideman. -- 1st ed.

ISBN 978-0-9973646-0-6 Print Edition
ISBN 978-0-9973646-1-3 Ebook Edition

CONTENTS

A Note To Readers

This memoir represents a true account of my experiences as I remember them. However, to protect people's privacy, I've changed a few names. Further, to make the story easier for readers to follow, I've recreated some conversations based on my memory of the information exchanged, sentiments expressed, and personalities involved. Although I've presented those recreations in dialogue form, they're not intended to be precise records of what was said. As for the book as a whole, I understand that all memory is fallible, especially after several decades, perhaps all the more so when it comes to stressful events. My memory is no exception. Some who shared the experiences on these pages may recall them differently. This memoir is not intended to dispute their recollections, only to offer my own honest account in hope of adding another perspective to the historical view of the Vietnam War.

To keep this memoir as accurate as possible, I've relied heavily on transcripts of interviews I underwent in the 1970s, when I collaborated with an author to create a novel based on my experiences as a prisoner of war. In those interviews, I gave a thorough account of my memories when they were fresh, within two years of my release from prison. I believe those old transcripts make this book a more reliable source of information than if I had dredged up the past by relying only on what remains in my head these many years later.

I don't wish to minimize the sacrifices or dishonor the memories of fellow POWs who endured imprisonment, injury, or death in Vietnam. I do hope this book serves as a reminder that the soldiers who fight our wars are humans and typically not saints. I believe we can learn something both from their heroic deeds and their blunders, their successes and their failures—mine included. I offer you my story in that spirit.

– *Robert Wideman*

For my grandchildren

1

SHOOT-DOWN

North Vietnam's thick green coast and white beaches looked peaceful from ten thousand feet. Glancing back, I saw the aircraft carrier's wake 125 miles away, cutting a perfect white line through the sun-struck blue of the Tonkin Gulf. It was a cloudless day with such unlimited visibility I could almost see the past behind me, the future ahead.

The Navy had trained me not to look quite that far, but to stay on mission. Had I dared look into the past, I might have seen my wife, Patricia, her clear blue eyes and cloud of blonde hair glowing from the candles she liked to light at dinner. I might have tasted her lasagna or the Baked Alaska she insisted on making though I would have been happy with a hot fudge sundae. During our five months of marriage in Lemoore, California, Pat's cooking had added fifteen pounds to my small frame. When I boarded the U.S.S. Hancock in San Francisco, the other pilots teased me that if married life agreed with me any more I wouldn't fit in the cockpit. Four months had passed since I'd last seen her, but I could still hear Pat's giggles the night we broke one of the legs on our bed, wondering if the crash woke our neighbor in the downstairs apartment.

Had I spared a moment to consider the future, I would have seen myself flying something bigger than this Navy A-4 Skyhawk. My plan was to fly commercial jets, filled with passengers bound for business trips and vacations. If anyone had asked, I would have said my destination was just ahead: the easy life. I had already served three and a half years in the Navy. Just twenty-seven more combat days and I'd be back in America. My future seemed as clear as the perfect blue skies over Vietnam. Our flight spotted no enemy aircraft, no surface-to-air missiles, no anti-aircraft artillery. No enemy radar. I was almost home. I could smell the Baked Alaska.

I glanced at the clock in the control panel: 1:00. Another flight would start the next run at 2:30. This was all standard. It did not occur to me to question the clock-like regularity of our daily missions. The U.S. military was a precision machine. I was grateful such decisions were not up to me. In a few moments, I would fire my rockets on the assigned target and turn back for the carrier. Landing on that narrow floating island without overshooting the mark would be the scariest part of my day. Still, it too would be routine.

The date was May 6, 1967. I was 23 years old.

There were just two planes in our flight, two pilots, each alone in his separate machine, neither man uttering a word. The only sound was the thrum of our engines as we rolled in on a small barge heading up a waterway about two miles inland. We lined up on the target. I felt but did not hear a sound I had never heard before. A small metallic click. In that moment, my transparent view of the world changed forever.

My aircraft rolled uncontrollably to the left and plummeted. The stick was off to the left about 45 degrees and I tried to shove it back to center. It wouldn't budge. Relying on the muscle-memory of my training to keep me alive, I turned off the autopilot, dumped my ordnance from the wing racks, and pulled the emergency generator. No luck. At about eight thousand feet, I saw a large hut rushing toward me, getting bigger by the second. When I saw six thousand feet spin by on the altimeter, I thought, If I don't eject now, I'm going to die in that hut in a ball of flaming metal. If I did eject, I was going to come down on land, not in the ocean as I had hoped, but there was nothing I could do about that.

The plane must have been diving at more than 500 knots when I reached overhead to the top of the seat behind me and pulled the face curtain. I heard wind shriek as the canopy flung itself from the aircraft and the ejection seat shot me into the sky. It felt as if someone struck me in the head with a baseball bat, and I saw a flash of black and then stars, like Tom the cartoon cat clobbered by Jerry the mouse. I struggled to tuck my arms and legs against my body so they wouldn't flail in the wind, but G-forces fought to pry me limb from limb. I tumbled through the air, flashes of earth, then sky, then earth, then sky, winking in and out of sight between my legs. "Work, God, work!" I screamed at the unopened parachute. It seemed as if twenty seconds passed, though in reality it took fewer than two seconds for the chute to deploy. I didn't feel the hammer-blow I had expected. I didn't feel anything. Too pumped on adrenaline maybe.

I was still falling too quickly.

I looked up at the chute, streaming like a scarf in the wind. The risers were tangled, so I reached up and pulled them apart. Intense pain shot through my fingers. I noticed two of them were bloody and crooked. Broken. Looking up past them, I saw that at least the chute had opened completely. It still appeared too tiny to hold me, but my descent slowed to a survivable rate. "Thank God," I muttered, not sure anyone was listening. Unlike my mom, I didn't believe in God. My dad had instilled in me his skepticism of religion. Too bad. Praying might have at least given me some hope.

My descent slowed, but my thoughts still raced. I did a quick mental inventory: two arms, two feet, a head. I was alive! I looked out over the Tonkin Gulf, a tantalizing two miles away, yearning for the haven of those welcoming blue waves. That was where I needed to be for rescuers to have a chance of getting to me before the North Vietnamese. Still, for one brief but gratifying breath, trouble seemed distant. Floating high above the war, I held on to the illusion of peace for an instant longer, stunned to silence by the intimate hush of the air after the rude rumble of the plane.

What a hell of a place to be, I thought with awe and regret. Then, Poor Patricia! I was breaking the promise I had made to her the day she saw me off at the airport. Always flustered by women's tears, I didn't know how to respond to hers either. So I gave her the knee-jerk, "I'll be back," though I knew full well I couldn't guarantee any such thing. Now I could see the Navy's casualty assistance team knocking on the door to her parents' place in Cleveland, Patty answering, her stunned expression. We had only met six weeks before we got married, had spent almost as much of our marriage apart as together. Any minute now she might become a widow.

I pictured Mom lighting a candle in church, Dad cussing under his breath, my brother flying a helicopter through these very skies trying to find me. Richard had volunteered for the Army and had recently received his own orders for Vietnam.

I could not yet feel sorry for myself. I had yet to even feel fear. It was too much to take in: what I had always known was possible, but had never imagined, was now becoming a reality. Yes, prayer would have been nice, but action was the only thing I could rely on.

What to do next?

I needed to contact the other aircraft in our flight. I reached for my radio, fumbled it, dropped it. Good thing I was in North Vietnam. I had always known that if I crashed in South Vietnam I would have been captured by guerillas. We all knew gruesome stories of prisoners suffering evisceration, torture, and death at the hands of the Viet Cong. Therefore in South Vietnam I considered my .38 revolver essential survival gear. However, rumor had it that the North Vietnamese had the resources to capture prisoners alive and take them to Hanoi for questioning. I thought a pistol would only get me into more trouble in the North. The point is, when I flew missions over North Vietnam, I always replaced the gun with a second radio. My only other survival gear of note was a pencil flare and a knife. That's why, instead of watching in horror as my radio tumbled to earth, I carefully retrieved my spare and pulled the cord to activate it. The beeping was startling in the silence as it sent out a distress signal.

"Raven One, this is Raven Two. How do you read?" I shouted into the radio.

No one answered.

"Raven One, this is Raven Two. How do you read?"

Still nothing.

That's when I heard the first sound other than wind since I'd ejected: a high, thin whistle. I recognized it as the sound of a bullet whizzing past. Terror finally seeped in, curling me into a ball as I attempted to protect my precious brains and heart with my flimsy arms and legs. I did not find out until years later that by the time you hear a round of ammunition it has already passed, but I doubt that would have made me act more rationally.

I was still too high in the air to see who fired, but the next thing I heard was a sound I remembered from many a New Year's Eve back home: people beating spoons against pots and pans, screaming and yelling. Then I saw them. It looked like a thousand people running across the rice paddies toward me, though it was probably more like two hundred. In training, I had learned that sounds start to become audible within five hundred feet. So when I heard the villagers, I knew it was time to prepare to land.

I had never done this before.

I prepared to be skewered as I skimmed along the tops of the trees, but luck was on my side; I whizzed past them unharmed. Then a shack rushed toward me, and I braced myself to crash in a tumble of wood and broken bones. But I missed that, too. I then noticed that the ground below was not only rushing up to meet me but was also rushing past me. A stiff coastal wind was pushing me inland, and I was moving horizontally at high speed, maybe 20 knots or so. What's more, I was coming in backwards. Our trainers had told us that landing sideways was best. If that was true, then this was all wrong. This was going to be rough.

I hit the ground so hard I bounced on my rear, sending an earthquake of pain vibrating through my back and neck, which had already received a beating from the cockpit when I ejected. When I came to rest, my chute was still billowing. Not dragging me but pinning me down like a bug, like a target. My broken fingers fumbled to release the lines as I spotted a Vietnamese soldier some fifty feet away. I could tell he was a soldier from his olive-drab shirt and trousers and his weatherworn pith helmet with the gold star—all suggestive of a gun I couldn't yet see. I wondered why he didn't step closer to either kill or help me. Then the scene came into focus: he was spreading his arms and shouting something to the dozens of villagers crowding him, all talking and yelling at once, pushing him back, pointing at me.

I scanned the crowd for weapons, but didn't see any of note. I continued trying to undo my parachute harness with shaking hands, avoiding sudden moves, looking from my task to the soldier, trying to communicate with my eyes: I'm only trying clear the lines, I'm not reaching for a gun. He looked terrified enough for both of us, scared of both the mob and me. Now I knew what the officer at Bunker Hill meant when he said, "Don't fire until you see the whites of their eyes." I had never before seen so much white around any man's eyes.

I'd never before seen a Vietnamese man up close.

I looked around for an escape route, but all I saw were small rice paddies, leading to raised dirt walkways, leading to more rice paddies. All of it enemy territory. Where would I go? Nowhere. I held up my one free hand and showed the soldier it was empty. As if that were the signal he'd been waiting for, he dropped his arms and the villagers rushed forward.

A few of them brandished knives, and I thought they might butcher me right then and there. Instead they cut me out of my gear, ignoring the zippers on my boots and flight suit, the releases for my parachute, the buckle on my helmet strap. I wondered if they did this because it was faster or because they'd never seen a zipper, buckle, or snap. I didn't dare move.

Despite the assault, my awareness of my surroundings grew. The air was hot, which I had expected, but also dry, which I had not expected. It was monsoon season. Where was the humidity, the rain? I had landed half-in, half-out of a three-foot-deep muddy trench. Great place for mosquitoes to breed, I thought. If the villagers didn't kill me, maybe some tropical disease would do the job. Soon I was sitting in that mud, stripped down to my white t-shirt and boxer shorts, with a couple of villagers on each side of me holding my arms and legs.

One kid who appeared no older than 14 reached into the chest pocket of my t-shirt, where I had a pack of Marlboros, a ten-dollar bill, and a packet of morphine syrettes. It dawned on me that was the same pocket where I stashed my Geneva Convention card and ID. If I lost those things, I could be tried as a spy. I slapped his hand away. He leapt back, eyes wide with outrage or terror, or both. That's when I noticed he was holding an old bolt-action rifle fixed with a French quatrefoil bayonet, a throwback from World War One. He jabbed it in my direction, and it occurred to me that, archaic or not, the sharp point could still kill me right quick. Instead, the boy rested it delicately on my navel.

Our eyes locked over the rifle barrel, his smug with power, mine unblinking. Christ, this kid doesn't have a clue what's going on, and if he sticks me or pulls that trigger it's all over. With that thought, I rediscovered the calm I'd felt earlier. It was odd to consider the possibility of death from such close proximity. How had I joined a war without thinking about death? I looked down at the mud caking my boxers. What an unglorified way to go.

An old farmer with a white scraggly goatee noticed the boy poised to skewer me. He must have been at least 70, but he wordlessly shoved the boy aside and took the rifle. I had no idea why he saved my hide, but I gave him a grateful look. My relief was short-lived. My captors weren't through with me yet. They blindfolded me and tied my hands behind my back.

I heard the roar of small jets approaching. The Rescue Combat Air Patrol was coming for me. I knew it was pointless. I was captured and that was the end of it. If I had landed in the gulf, rescue might have been possible. As it was, I wished I could warn them to turn around and forget me. It seemed to me that the only thing worse than being captured would be somebody dying trying to rescue me.

My captors let go of my arms and legs, and the bayonet kid shoved me deeper into the trench. The sound of the approaching engines was almost upon us. I slipped my blindfold up a bit and looked up to see an A-4 banking close overhead, maybe eight hundred feet. Surely it was a buddy from my squadron, but no telling who without completely ripping away my blindfold. The kid retrieved his rifle from the old man and popped several shots toward the belly of the plane even though at this distance his weapon was little better than a peashooter. Surely he was just firing for kicks. The problem was, he wasn't alone.

That plane, and several that followed, took a surprising amount of anti-aircraft fire from what had until a moment ago appeared to be nothing but a poor, idyllic farming community armed with pots and pans, a few knives, and a pea-shooting rifle. I heard 37- and 57 millimeter fire, as well as all kinds of machine guns scat tered through trenches around me. Small arms we called them, but they didn't sound small now that I was on the ground. I didn't know until that moment how loud they could be, how loud war could be. I'd been flying missions for eleven months, but I'd only heard war from the sky, where most of the sounds never reached the pilots. I had seen the results when our rockets hit targets, but the explosions were faint, distant, less noisy even than watching a World War Two movie in a theatre, as if they were happening in a place and time that bore no relationship to me.

I hadn't even heard my own plane crash. Where did it go down? I wondered.

Here on the ground, this was another war. More real, visceral. I could hear the boom and roar of battle and the shouts and screams of villagers, feel the ground rattle and my ears compress, see three-dimensional humans running and jumping into trenches like the one I was in. I now shared a strange part in the fate of these people, these Vietnamese we had been taught to hate, these men we called "gooks." Might I have survived the crash, the angry mob, and the bayonet kid, only to get killed by friendly fire?

The firefight went on and on. All over one downed pilot. I had heard that this tiny country of so-called ignorant communists was losing the war, but that's not what it sounded like from my new position. The enemy was more formidable than I'd imagined.

My captors, now down to about a half-dozen guys, yanked me to my feet and pushed me in front of them. They shouted something in Vietnamese that I didn't understand, but I understood their gestures: Get moving! That was easier said than done. It was only then that I felt the toll taken on my body, first by my ejection from the plane, then my rough landing, then the manhandling by the villagers. My back and neck now felt as if someone had hit me with a concrete sledgehammer. I could barely shuffle my feet without pain flooding every muscle.

We walked inside the trench to evade the worst of the air assault that continued unabated. I only took about a half-dozen steps, feet sinking through the thick mud until it sucked off my socks. Barefoot now, wearing only my underwear, peering under my blindfold at the muck between my toes, I stumbled forward. I had no idea where we were going, or what might happen to me when we got there.

2

YOU DIE

The world looked hazy around the edges of my blindfold: shifting glimpses of smoke, sky, and mud. Listening didn't help, just lost me in Vietnamese conversations I could not understand. I felt sore to the bone, and it took everything I had to put one foot in front of the other. We rose out of the trenches and into a small thicket of trees. After about a quarter of a mile, I was pushed through a doorway. I ducked inside what seemed like a small bamboo hut, where something thumped my face repeatedly, like a small bird flying into my head over and over. It didn't hurt, but it was disconcerting. I peered through a gap in my blindfold and saw a pretty girl of about 16 sitting near me, wearing a thick black braid and black pajamas, reaching up to hit me in the head with her rubber sandal.

She punctuated each hit with an angry rush of Vietnamese. I had no idea what she was saying. Maybe, "That's for the Vietnamese your bombs killed!" *Whack!* "And that's for my neighbors your plane crash killed!" *Whack!* "And that's for interrupting my work!" *Whack!* The rubber sandal seemed almost silly as weapons go, given the deadly gunfire raging outside, but the persistence of her hatred was impressive.

Someone tugged on my elbow and prodded me downward until I was lying on the hut's dirt floor. Since my hands were tied, this was a clumsy affair. I had no idea how many people were in the room though I sensed them nearby. It occurred to me we were waiting for the air assault to pass.

After some forty minutes, the gunfire died down, and my ears rang in the new silence. I fought a fleeting feeling of abandonment, though I'd known from the moment I hit the ground that I had no hope of rescue. My countrymen were gone. I was alone in the dark isolation created by my blindfold, lying on the floor of a hostile country.

A pair of hands pulled me to my feet, untied me, and removed the blindfold. I blinked into the dimness of the hut. About a half-dozen young soldiers stared up from the floor with varying degrees of curiosity. An officer of about 35 wearing a pith helmet with a gold star sat on a bed next to a small table. He thrust into my hands a small square of paper with the following words typed in English: *Name, Rank, Serial Number, Date of Birth, Unit.* He then handed me a cigarette and pointed at a pencil on the table. It seemed I would be allowed to smoke if I gave the information requested on the form.

As simply as if I'd been through this a hundred times, I picked up the pencil and wrote in the first four blanks: Robert Wideman, Lieutenant Junior Grade, 689953, 10/17/43. I did not write down the name of my unit, feeling a momentary pride that I was able to remember the military code of conduct, which stated that as a prisoner I only had to give captors my name, rank, serial number, and birthdate. I handed the paper back, also gratified to notice that my hand was not shaking. He scanned the paper. I had no idea whether he knew what it said, but he gave a satisfied nod, pulled out a lighter, and pointed at the cigarette I still held.

I had quit smoking two months earlier, but didn't hesitate to hold the cigarette between my lips and lean in toward the lighter, my hands shaking slightly in anticipation of that calming first drag. It seemed the sociable thing to do, and playing the easygoing prisoner seemed to weigh in favor of my survival. On the other hand, maybe this was the old-fashioned offer of a last cigarette before my execution. If so, that seemed another good reason to take him up on it. I inhaled deeply, careful to turn my head away from him and the others before exhaling a small cloud of smoke into the confined space.

I sat on the floor to finish the cigarette and felt surprised to see the girl who had smacked me with the sandal now holding out a tiny cup and saucer. I accepted them, and she poured tea into the cup from a small jug. The cup was no bigger than a demitasse, and I knocked it back in one gulp. I had never been so thirsty in my life. She poured another, which I knocked back just as quickly. I kept drinking and she kept pouring, with the same steady diligence she had used in hitting me with her sandal. Her expression was neither friendly nor unfriendly. I guess mine wasn't much different—how could either of us feel comfortable with one another in such a situation?—but after I finished my sixth cup or so and handed it back, I think I muttered, "Thank you." She nodded and left.

That's when I noticed the voices in the village growing louder, closer. One of the guards hauled me to my feet and pushed me toward the door. Maybe it was drinking all that tea, or maybe it was the fear of facing a mob, but the blood ran to my feet and my head felt light as a balloon. My vision turned yellow, then white, then blank. I stumbled forward, but even though I was no longer blindfolded I could not see. I shut my eyes, willing myself to stay conscious. I imagined what a clumsy, filthy, broken mess I must look like to the people outside the hut, who were a blur of voices.

I opened my eyes again and they came into focus, maybe twenty or thirty people, mostly old men, women, and children.

We walked only a few huts down before they shuffled me into another hut. I wondered at this, until I heard more planes approaching. This time I heard propellers: definitely A-1 bombers, several of them, not cruising at the 300 to 350 knots I was used to, but putt-putting along at 150 so they could drop their payloads. Oh, my God, I thought, those guys are gonna get hammered. I heard them motoring around and around the village for some fifteen minutes, heard every gun within maybe five miles open up on them: *bum-bum-bum-bum-bum.* We were right in the middle of it.

Though I was pinned down with the North Vietnamese, in my head I was still a pilot and my thoughts were not on the ground but in the sky. Jesus God, you poor guys, give it up. It's over. I'm done. This time the bombing and gunfire subsided in about fifteen minutes.

We soon resumed walking out of the village, past the small green shoots in the muddy rice paddies, west into the sun, toward the foothills. The angle of the sun suggested it was three o'clock. Just two hours since I had fallen. It felt like a lifetime.

I figured if I could survive the night, I might be a prisoner for a few months at most before the United States won the war, as it always did. A second look at my captors bolstered my hopes. Sure I was half-naked, but my captors looked far from parade-field proud: one wore a military shirt and helmet but shabby farmer's pants, another wore khaki pants and some sort of t-shirt, another wore what appeared to be civilian clothes but in camouflage colors clearly meant for combat. There was no uniformity to their arsenal of side arms and rifles: a hodgepodge ranging from a Springfield to an AK-47. I did not reflect on the images this conjured from my school-day history books: images of underdressed, underfed, Colonial American rebels with scrounged supplies defeating the spit-and-polish Redcoats.

As we left the dirt roads and hit the main road, my captors prodded me on with their hands and rifle butts, urging me to hurry, "*Di di mau, di di mau*" (*Quickly, quickly*). They pointed behind us, and I turned to see some two to three hundred villagers surging toward us across the rice paddies like a human tide, half a mile away and closing in quickly. My captors gave me knowing nods, their meaning clear: If those people catch up to you, you're toast! It made sense. Just two hours before, I had been sitting in my cockpit watching a hut speeding toward me. My A-4 must have crashed into a village, killing who knew how many civilians as it exploded into flames. I struggled to move faster, and I resisted the urge to look back. My top speed wasn't much. I must have pulled every last muscle jumping from that plane and I was hobbling like an arthritic eighty-year-old.

It was impossible to avoid the wrath of at least a few Vietnamese. We passed through a half-dozen villages, and in each the locals swarmed us, men and women, young and old. Many of them threw sticks and stones at me or tried to punch, slap, and kick me as I passed. I bowed my head and shut my eyes, lest someone furiously gouge one out. My number one goal was to make it to Hanoi in one piece.

I was amazed by the unflinching professionalism of the guards in the face of the mob. They closed ranks and crisscrossed their rifles over my head, using their own bodies and firearms to take the brunt of the attack. I had no illusion that they did this out of sympathy, but rather to protect a valuable asset: I was now a potential source of information, a bargaining chip, human leverage. Of course, they were not about to sacrifice themselves any more than necessary. It was obvious they purposely let a few whacks make it through their barrier.

Meanwhile, the American air strikes continued to come and go, though by now the attacks had nothing to do with me. When each raid approached, the guards would force me down into the trench alongside the road, then jump in after me, where together we would watch two to four planes come in over the beach, drop their bombs inland, and head back for the gulf. During mission briefings aboard the carrier, I had received the impression we were turning up the heat on this country. But from my new vantage point, it seemed the locals swatted at our aircraft as if they were little more than flies.

Whenever we stopped to duck an air raid or to rest, I lay down. Each time that the guards ordered me back on my feet, I wondered if I would make it. It took no less than five minutes to force myself from standing to prone or from prone to standing.

We stopped in one village long enough to quench our thirst, and this time the villagers stopped hitting me long enough to stare. Just stare. It seemed that not only did they want a swing at me, but also wished to gape at the foreigner, the white man, the American imperialist. I tried not to make eye contact. I felt like an animal in a zoo.

We moved just uphill from that village and parked ourselves in a trench some fifty yards off a broad dirt roadway. Fifty or so villagers sat around us, murmuring among themselves. A young man lit a cigarette for himself and then, after checking with my guards, stepped up and gave me one. Again I wondered if this cigarette might be my last. Not that any of the guards had shown me animosity, but I was clearly an inconvenience. Who would bother to argue with them if they said I died attempting to escape, got killed in an American attack, or fell into the hands of an angry mob?

After I finished my smoke, one of the guards produced my flight suit, boots, and helmet, and gestured to me to get dressed. I put everything on as fast as I could, which was excruciatingly slow. When I finished, I looked up to see a young man aiming a Rolex camera at me. I was pretty happy to see the camera because I knew that my squadron had no way of knowing I had not been killed. I hoped that a photo of me would make its way to American eyes so that someone could tell my family I had been captured alive.

I wanted to look strong in case my family or my squadron might see the photo: a poster-boy for military morale, a happy and healthy son to ease any mother's fears, a handsome young husband a wife would consider worth waiting for. With that in mind, I made the mistake of taking a macho military stance and smiling. Several guards and villagers scowled and raised their voices in outrage. Perhaps they thought it was evidence that Americans were a sick people, smiling at the misfortune of families into whom they flew their planes, or acting with outsized conceit even in the face of defeat. Their reaction prompted me to erase all traces of cheer from my face, thinking, Maybe that's not such a good idea if I want to stay alive. I kept my expression stoic, as the photographer snapped a picture or two.

The guards then motioned for me to sit, and they put on their helmets and sat around me holding their rifles. One of them beckoned to someone nearby, and I was surprised to see the rubber-sandal girl step forward. A guard handed her his rifle and she joined them, striking a pose for the camera. I didn't dare turn to see if any of them smiled for this portrait, but I doubted it. This was a capture scene, maybe for the world to see, and they seemed earnest about their roles as representatives of Vietnam.

After the photo shoot, they forced me to strip down to my skivvies again.

The sun began to drop behind the mountains, casting a golden glow over the fields of the delta below, and I became just another face on the hill. Together we watched the planes come and go and listened to the bombs fall on what I had previously considered strategic targets. From this angle, it all looked like little more than simple farm country.

Someone lit a cooking fire nearby, and soon the rubber-sandal girl brought me a bowl heaped with rice, a potato, some weedy-looking greens, and a fish head. I must have been starving, but the food aroused almost no response in me. My stomach did not rumble with anticipation. I did not salivate. The fish head on top didn't help my appetite. Still, I forced myself to eat, knowing it was important to stay strong.

The girl had also handed me a pair of chopsticks. Normally I was pretty adept with those, but I'm right-handed and that was the hand with the broken fingers, which were very swollen by that point. In fact, both of my hands were stiff. When I tried to use my left hand, the chopsticks kept tangling up so that I could barely scoop the rice into my mouth. It grew very quiet, and I looked up at the guards perched on their haunches around me, their own bowls untouched, staring at my lack of progress. They all burst into laughter. One of them mimicked me with his chopsticks, and they laughed harder. Finally the sandal girl reached into the circle, removed the chopsticks from my hand, and handed me a large aluminum spoon.

Although the food was bland, I did feel better. Clearer. Almost hopeful. After the meal, the sandal girl handed me a tiny cup of lukewarm yellowish liquid. I sniffed it, but it had almost no smell. When I drank it, it tasted like water, though I believe it was weak tea.

As I drank the tea, a little boy around eight years old broke away from one of the nearby families and ran around above me, along the lip of the trench, giggling.

Every now and then he stopped right in front of me, grinned, pointed an index finger in my face, and said, "Ah! You die! You die!"

I felt something catch in my chest at the real possibility in his words. But I only smiled back, pointed at my chest, and shook my head, "Me? Nah. No die."

He shook his head hard. "You die!"

"Oh no, not me," I repeated. Just who was I trying to convince?

After sunset, the boy and his family returned downhill with the rest of the villagers and three of my guards. My remaining three guards and I climbed out of the trench and stood in a clearing next to the road, obviously waiting for something—transport I assumed.

A guard offered me another cigarette. Why not? What else was there to do? I blew smoke into the day's last orange light and thought about my squadron on the carrier. At that moment, most of them were talking, smoking, and laughing after dinner, getting ready to climb into their bunks and catch some shut-eye. Some were doing their final flight checks for night runs. Tonight I would not be joining any of them. I would never join them again. I was trapped in Vietnam until either the war ended or I did.

I flipped through random images from my life: my wife's blonde hair shining in the sun, my little brother and I laughing in the woods near our house, my mom holding my hand as I crossed a street, my dad and I sitting next to the radio listening to *The Shadow*. I felt frustrated as I tried to recall who *The Shadow* was. What was his alter ego's name? Damn, it was on the tip of my tongue. The question jumped across my brain like enemy flak: What's his name, what's his name, what's his name? Let me at least remember who *The Shadow* is before I die!

Darkness fell and the road lit up, as if someone had flipped the "on" switch for the world's longest conveyer belt. On the road in front of us, snaking from north to south, a line of trucks materialized from the thick air, one every hundred to hundred-fifty yards apart—for as far as my eyes could see. They were all crawling south at a slow but steady five to ten miles an hour. On every truck, both headlights were punched out, replaced with a single light attached to the front bumper, and over that a bamboo cover to ensure the light only shined straight down at the road. Few, if any, pilots would be able to detect those lights from above. I had never seen so many trucks during my night missions. Yet it was clearly not the first night for this supply line.

Whenever the sound of jet engines approached, the trucks halted and the lights shut off, completing their illusion of invisibility on the Ho Chi Min Trail.

I recognized the pattern of the American flights as they came and went, because I had flown the same pattern, but I had never before considered it with the sight of these trucks before me. During the day, two to four airplanes would hit the trail about once every hour and a half. At night, the interval was about the same, but only two planes would make each run; one plane to carry bombs, the other to drop flares for better visibility. Those flares also meant better visibility for the North Vietnamese Army (NVA). I watched in horror as small orange balls of light exploded amid the slowly falling white light of the flares. The orange balls were artillery tracers.

Within five minutes, the planes were gone. Hundreds of trucks turned their lights back on and lurched forward again as one.

I had flown those missions like clockwork. Like clockwork. And the Vietnamese had all but set their watches to it. The truck drivers knew that all they had to do was stop for five minutes every hour and a half. The rest of the night they just...kept...driving. Patiently, slowly, tirelessly.

I felt stricken. In February, barely three months earlier, we had waited through a weeklong stand-down to see if the North Vietnamese were ready to negotiate. During the stand-down, an admiral had visited our ship and given us an intelligence briefing. I remember him saying, "We've stopped ninety-eight percent of the supplies going from North Vietnam to South Vietnam." My fellow pilots and I had cheered, congratulating ourselves on our great success.

But on this night, I watched that endless line of trucks continue its relentless push south and knew the admiral was wrong. I thought, The Navy doesn't know. I have to find a way to get a message through, to let our guys know that our attempt to stop the supplies has failed.

Years later, I took a course in American legal history with Dr. Kermit Hall, who had worked in the U.S. Army Intelligence Corps during the Vietnam War. He told me he had flown over the Ho Chi Min Trail many times throughout 1968 and '69. He said the U.S. would have had to fly twelve hundred sorties every day to stop the supplies heading down that trail. He said we could only put up three hundred sorties a day over the Ho Chi Min Trail. We had zero chance of stopping those supplies, and the admiral who had briefed us surely knew.

Although I did not have access to Hall's information that night, the giant caravan that rumbled past me in the dark told its own story. I realized we would be lucky to stop one-tenth of one percent of Vietnam's military supplies. I made eye contact with one of the soldiers who guarded me. His eyes seemed to reply, You poor, stupid bastard.

I thought of senior military officers back in the States telling me that North Vietnam could not keep this war up for two more years, my superior officers reporting to us that we were beating the NVA to the point that "we just don't see how they can hold up."

Through most of the long day after my crash, I had clung to my belief in our political leaders, in military intelligence, and in the hope that I would not be a prisoner for long. Surely, America would win this war in a couple of months. Then I would be free to return home to my family. But that night, my naiveté became obvious. I breathed in the exhaust of the trucks for the next two hours and realized I was going to be in Vietnam for a very, very long time.

I thought of the boy who had taunted me, "You die!" Perhaps he was right.

3

GOING NOWHERE

Hours passed, but none of the passing trucks stopped for us. My two guards led me back uphill, where a young couple strolled up and joined us. I could almost imagine we were at a Fourth of July fireworks show back home, watching the orange artillery tracers and white flares flash across the sky, listening to the echos of some other war. The couple and the soldiers talked and laughed while I remained aloof. I was not one of them and it didn't seem fitting to pretend otherwise, but it was difficult not to smile and nod along with their good humor. A guard caught my eye and pointed from the girl to me with raised eyebrows, hinting, Maybe you want her to be your girlfriend too? She and the boy giggled at my confusion as I struggled to respond without offending. I held up a hand, "Not me. I'm married," and chuckled softly—too much and they might think I was making fun of Vietnamese women, too little and they might think I was disgusted by Vietnamese women. They laughed, and that relaxed me a bit.

Soon the guards moved me to a barren tree near the trench, and sat me facing uphill instead of down toward the road. When I looked over my shoulder to see what was going on, they snapped, "No!" So I turned back around and stayed that way until one of the guards walked up to me, gave me a nudge, and motioned for me to stand. The young couple was gone, but a third soldier had joined us.

The four of us walked downhill through some brush, crossed a bridge over a creek, and emerged by the roadside, where a large truck idled. The truck was painted in camouflage colors, with jungle foliage and pine boughs scattered atop sandbags piled high around the truck bed. The fresh scent of pine mixed with the nauseating stench of diesel. The all-night convoy of supply trucks continued past.

Three soldiers stared down at me from the truck bed, and my guards motioned for me to join them. I felt so stiff and sore that climbing up into that truck bed seemed akin to summiting Mount Kilimanjaro. A guard gave me a swift shove from behind and leapt up to join me. We entered the parade of trucks headed south on the dirt road. This surprised me because Hanoi was due north and I knew that's where the NVA took captured pilots. I wondered what they were going to do with me if they did not take me to the Hanoi prison camps.

Someone gave me my sixth or seventh unfiltered cigarette. I accepted that I was now a smoker for the duration of my imprisonment, or possibly for the rest of my life.

We puttered along for half an hour before I got out with a handful of guards. They loaded my ejection seat onto my shoulders—complete with parachute, helmet, and oxygen mask—and secured all fifty pounds with my torso harness. Then we walked east on a dirt road. I was still barefoot, wearing nothing but my drawers. Within minutes I was lagging. They prodded and scolded me to hurry, but my battered body couldn't take the load. One of them grabbed the ejection seat in irritation, threw it on his back, and lugged it for me. It was maybe 10:00 p.m., nine hours since my plane had gone down. I sensed trees around me, but didn't notice much else about my surroundings, or about myself for that matter: neither fear nor pain nor discomfort. Nothing but a bottomless fatigue.

We stopped at several checkpoints. Each time, a sentry called out a challenge and one of my guards muttered something that sounded like a password. As my eyes adjusted to the dark, I made out silhouettes of helmets, heads, and machine-gun barrels rising from under the bushes next to the road. These posts surprised me. I'd been told so often about the backwardness of these people that I didn't anticipate such professional organization. We believed we were winning the war in Vietnam because we assumed the Vietnamese were ignorant, poor, and disorganized, unable to win against our superior American war machine. Now I wondered: Were we even winning?

Each checkpoint was followed by a village, where voices and music crackled from a loudspeaker overhead. Since the villages were only about half a mile apart, the speakers created a stereo effect. I had heard that people in third-world communist countries had no radios, that they were brainwashed by anti-American propaganda shouted over such speakers. I accepted that, but I was still curious about what they were saying.

We must have walked ten miles. The entire night felt like a hallucination. I was on the verge of forgetting who or where I was. I only knew I was thirsty and needed to pee, but the guards would not let me drink or relieve myself for the duration of the walk. The pressure in my bladder became my whole world.

At about 2:00 a.m., we stopped at a large concrete building, its whitewash faintly illuminated by whatever scant moonlight penetrated the smoky skies that night. The guards let me pee in the dirt outside; I was surprised I'd made it so far without wetting myself. We stepped into a large empty room with glassless windows and an earthen floor that rose to a low mound in the center. My guards motioned me toward the mound, where I collapsed.

It felt like I slept fifteen hours, but it was really more like thirty minutes. When I awoke, I felt the most desperate thirst I'd experienced in my life. I imagined my organs withering inside me. The silhouettes of other people in the room wavered before my burning eyes in the dark, so I croaked in their direction, "Water!" I pointed at my throat. "Water!"

Though nobody had responded to any of my earlier words, it seemed they understood *water*, or at least why a man who'd walked all night might point at his throat. An old woman wobbled away and returned with an archaic, cylindrical, blue-and-white porcelain jug with a bamboo-and-rope handle. She held it out to me, no cup. I clutched it and lifted the spout to my mouth, wincing at the pain in my neck as I tipped my head back to gulp as much water as I could, not knowing how long it might be until my next drink. The room came into focus, revealing two or three Vietnamese families: perhaps a dozen children, young women, and old folks.

I reminded myself that tourists paid good money to travel to distant lands and experience other cultures, so part of me was intent on soaking up this strange new atmosphere. The red tip of a lit cigarette revealed guards in the corner. Someone handed me a cigarette. The smoke seared my throat, but soon sent a calming pulse through me. The others crouched on their heels as they smoked and talked, so I tried to do the same, but by the time I finished the cigarette my legs were cramping, so I sank back to the ground, depleted. I couldn't clearly make out anyone's expression in the dark, though it seemed we stared at one another with mutual curiosity. They muttered among themselves, but their faces remained focused on me as I dozed off.

Two minutes or two hours later, I was jostled awake and propelled outside to a jeep. I felt relieved I didn't have to walk any more, though I worried a bit when the jeep continued carrying me farther south. Prisoners were tortured and killed in the South, but it didn't make sense to me that NVA soldiers would travel so far out of their way to torture and kill me. They could just as easily shoot me here if that was their aim, couldn't they?

We stopped a couple of hours later. This time only one guard walked me through the darkest part of the morning. I felt certain I was going to faint dead away, and twice I did sink to the ground like a rag doll, eyes closed. The guard didn't disturb me in my slumber, just left me slumped there for a couple of minutes before hauling me to my feet. He was probably exhausted too. Soon a new set of loudspeakers lured me into the next village on a river of soothing music.

We stepped inside a hut, where I slumped against a pillar. Somebody raised a lantern toward my face. When the red light hit my eyelids, I opened them slightly and croaked, "Water. Water." I heard a man's voice speaking in urgent tones. A woman whom I assumed to be his wife scurried into the room, took one look at me, and gasped in such a way that if I had the energy for it I might have felt sorry for myself, imagining the absolute fright I must look. I suppose it was she who brought me water. I have no idea. I only remember a sensation of wetness in my mouth and throat. Then I was gone, a "Nowhere Man" as the Beatles might have said, until sunrise on May 7, 1967, the dawn of the day after my plane went down.

I felt a bamboo mat beneath me, but had no idea how it or I had gotten there. I heard a crackling noise that coaxed me to open my eyes, but when I tried to turn my head I moaned. My neck felt like a block of concrete threaded with crushed nerves, and I had to turn my shoulders to get my head to turn. I saw the man from the night before cooking something over a fire in the corner.

Soon I heard people outside: adult feet shuffling between tasks, children's feet running from place to place, young people shouting, old people murmuring, bursts of laughter. Knowing that someone was likely to walk in at any moment, I spent some twenty minutes struggling to wake and push myself to a sitting position. The wife walked in, cast a glance at me, then whispered to her husband. He turned an exasperated face my way, left the fire, and approached me. He poked me gently but firmly in the shoulder to get my attention, and then pointed to my genitals, which were hanging out the crotch of my skivvies. He frowned and nodded toward his wife at the fire.

I tucked my privates out of sight and dipped my head in apology. "Sorry."

He turned away, muttering and shaking his head.

Once I was decent, the woman brought me a meal that looked much the same as the meal of the night before: a bowl of rice with a few weeds thrown on top, followed by hot tea the color of piss and probably only slightly better-tasting. I supposed I was getting the dregs of leftover tea that might otherwise be thrown away, or maybe it was what they all drank.

After breakfast, a young soldier came in and handed me a laminated card printed with phrases in several languages, most notably English and Vietnamese. The phrases were grouped under photos indicating transportation, medical issues, bodily needs, and other subjects a prisoner and captor might need to discuss. In the military, we called this sort of translation card a *pointie-talkie*. The soldier who handed it to me indicated that I could use it to ask him questions.

I had grown apprehensive about the many hours I had traveled south away from Hanoi, the city where I had expected to end up in an official prison camp with other Americans. So I first pointed to *Where am I going?*

The pointie-talkie has its limits. It took me pointing at the same line and him repeating his multisyllable answer several times before I made out a word that sounded like "Anuh."

"Hanoi?" I over-enunciated the word.

He nodded, impatient. "Anuh!"

Still not sure we were on the same page, I pointed at the question, *How Far?*

"One-hunre-fitty mie," he said, holding up fingers for emphasis. One hundred fifty miles sounded about right for Hanoi.

On the other hand, now that I could barely walk, I felt worried about the distance. So the next question I pointed to was something like, *How will we travel?*

He pointed to a drawing of a person who was obviously walking and burst out laughing as if he thought this was funnier than hell.

I tried to chuckle, just to be agreeable, but shook my head and pointed at my legs. "No can walk that far, buddy."

He laughed harder, nodding and pointing at the walking cartoon with increased energy.

Moving on, I showed him my two broken fingers and pointed to a phrase that more or less said, *Do you have a doctor?*

He said nothing, but exited the hut.

He soon returned with what appeared to be a little glass tube of clear medicine. He snapped off one end with a knife and handed it to me. I stared at it, thinking, This is ridiculous, my fingers need to be splinted. What's this stuff supposed to do? When I hesitated, he mimed stabbing the broken end into my finger. That seemed even more ridiculous. Cut myself with the glass? What was in it? What kind of medical training did he have? I played stupid, giving him a blank look that suggested, I have no idea what you're talking about. He grabbed the broken tube from me and made the same stabbing motion, but stopped short of jabbing the glass into my hand and instead shook the liquid onto the wounds. My fingers were red and raw, so I could see why he might think disinfectant would help, if that was what it was. But it smelled and felt like water, and might as well have been water for all the good it did. My fingers remained unset and throbbing.

That was the end of pointie-talkie, which was about as useful as that little vial of water.

After that, a couple of guards had me put on all my gear again and took me outside for more photos. Then they stripped away my clothing again and left me in the hut for the midday siesta that seemed to be the norm in every village. I tried to nap, but kids kept running up to peer under the gap in the bamboo wall and throw pebbles at me. Most were small tots with poor aim, but a few of the older kids, maybe 12 or 13 years old, clipped me pretty good. I glanced around to make sure nobody was in my vicinity, picked up the offending pebbles and fired them back at the faces laughing at me under the wall. The kids scattered, laughing even harder.

When they left, something else hit me: it was yesterday evening in the States. My wife must have gotten the news by now. What did they tell her? Was I listed as captured, missing, or killed? I thought of the last time I held her, in an airport amid strangers, just another soldier and his girl pretending to know they'd see each other again. What would she do if she heard I was dead? How long before I saw her again? I pictured coming home soon and kissing her like that sailor kissed the nurse in Times Square on V-J Day. Or years later, skeletal, gray-haired, weak—my bride nowhere in sight. Or maybe I was the one who would not be there, dead from any number of macabre prisoner-of-war scenarios: shot in the head, beaten to death, starved until I was nothing but a ghost. Gone.

One minute passed, then another. Cut off from all I knew or understood, I floated between boredom, loneliness, and despair. Most painful was my lingering hope: maybe the U.S. and Vietnam had signed a peace treaty, even now the war might be over. But that infinite line of trucks heading south told me the war might never end. Why think about it? There was nothing I could do. Another minute passed. How to shut my mind off? I replayed the crash, wondered what that click in the cockpit was, what brought my plane down. One minute I was flying, the next I wasn't. A one-in-a-thousand chance. Rotten luck, I told myself, and then, You're one lucky dude to be alive. Another minute passed.

I longed to hear the airy lilt of Patricia's voice, worried I would forget it. We had so little time together before I left. This feeling growing inside me went beyond homesickness. I knew no one here, had not a soul to talk to or laugh with, found nothing familiar in the people or landscape that my brain could latch onto. I was alone. I'll admit I shed a tear or two. Then the kids returned, crouched on their knees, staring under the wall, elbowing each other and giggling. I wiped my eyes and composed myself. I didn't want the Vietnamese to see me as weak, not even their children.

There was nothing I could do but accept my situation. I stared at the thatched ceiling, seeing nothing, trying to feel nothing.

After dinner, two guards led me through the village to a large open-sided, thatched shelter where at least two hundred people of all ages had gathered, the entire village it seemed. At my approach, the normal chatter that accompanies such a crowd fell to a murmur, and the high buzz of the cicadas in the surrounding trees took over. The guards made me sit on the concrete floor, just sit there, while everyone stared. I was a circus freak, the after-dinner entertainment in a town without TV.

One four- or five-year-old kid ran up to me, squealing, and pinched my love handles, hard. A few people chuckled indulgently at the adorable little bastard. Nobody stopped him. I would have loved to grab him by the throat and throw him down. Now *there* would be a circus act for them: lion eats child. Instead, I knocked his hands away and kept my eyes on the crowd, trying not to offend.

The crowd and I stared at each other for several minutes before the guards led me back to the hut.

Just before sundown, I walked out of the village, two guards in front, two behind. The entire village stood and sat in a crooked double line on the curving main path between the huts and the rice paddies, a gauntlet of sullen and curious stares. No slaps or stones, just silence.

At the end of the gauntlet, a large jeep with a canvas-covered bed waited. I sat on a board in back, and a guard wearing a t-shirt the color of dried blood sat on the board across from me, immediately pointing his assault weapon in my face. It was similar to an AK-47, but it had holes in the barrel. He kept it pointed at me as we drove away, finger on the trigger, arm hanging lazily at his side as if he were indifferent to whether or not he shot me. Many minutes later, I was still looking nervously down the dark bore of that barrel. What if we hit a bump in the road and his finger jerked, unintentionally pulling the trigger? What if I twitched unexpectedly and he reacted without thinking? This kid looked to be about 18. What did he know about firearms or the finality of firing one at another person?

My father was a hunter. I remember him standing behind me, the smell of his cigarette smoke blended into his soft flannel shirt, as he showed me how to properly hold a rifle when I was only four. "Son, never point a gun at anything unless you're going to shoot it." Years later, when I was about 14, my eleven-year-old brother, Rich, and I were shooting at birds in the trees outside our back window in East Aurora, New York. A few birds tumbled to earth in a flutter of feathers, but the rest of the bullets went wild. My dad pulled into the driveway, heard the shots, and charged into the house.

"What the hell are you doing?!" he roared, yanking the rifle from my now-slack grip.

"Just shootin' birds, Dad," I muttered.

"And what do you think happens to the bullets that don't hit the birds?"

I hung my head. I hadn't thought about that. The rest of those bullets might have rained down on our neighbors a mile away, like artillery fire. He blamed me, not Rich, because I was oldest. I was surprised he didn't beat me, as he often did when he was enraged. I suppose he was still overcoming his shock at the human tragedy I could have caused.

The young Vietnamese soldier sitting across from me in the jeep was only a few years older than the dumb kid I had been back then. After about forty-five minutes sweating out the death-threat bobbing lazily in my face, I'd had enough. Careful not to make sudden moves, I said, "Hello." His sullen expression didn't change. I slowly lifted my hand, laid it against the side of the muzzle, and nudged it aside, just enough so it no longer pointed at me. I gave him a half-conciliatory, half-pleading smile, and held up my hands in surrender, trying to pantomime the idea: You don't have to point that thing at me. I'm not going anywhere, and I'm not going to try to do you in. All he saw was my hand on the barrel. His eyes widened, his jaw clenched, and he jerked the muzzle back into my face as he looked me square in the eye. I got the message: Don't try that again, or maybe I will shoot you.

I didn't mess with his weapon again, just gripped the board under me so I wouldn't fall during the bumpy ride over muddy, rocky, potholed roads. My fingers grew sore from the effort to hang on. I knew the two broken ones would never be the same.

Every vertebra in my neck smashed into the next. What was up with my neck?

At least we were driving north this time. It seemed I was, indeed, headed to Hanoi. After dark, we passed the line of supply trucks heading south. It started to rain, and the soldiers guarding the truckloads wore ponchos. I was still mostly naked so I was glad to be listening to the pelting of the hard rain from under the protection of the canvas, though I was covered in filth and sweat and part of me longed for the cleansing feeling of water running over me.

We drove for several hours, until 1:00 or 2:00 a.m., stopping a few times for the driver to confer with the truck-drivers heading south. I was astonished to see him gesticulating and pointing at the roads with the kind of confusion that made it clear he was lost and asking for directions.

At one stop, someone came to the rear of the jeep, shined a flashlight in my face, and took over guarding me. At least Mr. Itchy Trigger Finger left. The new guy put a blindfold on me and tied my hands behind my back, so I knew we must be getting close to our destination—the prison camp, I hoped. For the next hour, it felt as if we were going in circles, but maybe that was just because I could no longer see. Several times I fell off the bench and banged my knees on the jeep bed because with my hands tied I could no longer hang on.

We stopped and one of the guards helped me down from the truck. "Can you see your feet?" he asked in English.

"Yes."

And away we walked. This time we only humped it for about thirty minutes, and it didn't seem so bad because I was filled with the hope that I was finally going to see other Americans.

4

HOGTIED

The guards led me blind and stumbling up another hill and into another hut. Someone sat me down, still tied and blindfolded, on a wooden board laid atop the damp ground. The first thing I noticed was a terrifying sound, guttural like a gurgling crocodile. It scared the shit out of me. Were they about to feed me to a wild beast? The voices of several young Vietnamese men erupted into laughter. Maybe they were preparing to play a joke at my expense: lunge at me with some hideous swamp creature the moment someone lifted my blindfold, another effort to scare the prisoner.

I noticed a pungent whiff of tobacco smoke with a hint of sweetness. It burned my throat, causing me to hack and cough. Was this some strange Vietnamese brand of cigarette? Did it have something to do with the animal I couldn't identify? It was unnerving, especially since no voices of other American prisoners materialized in the din as I had hoped. Still, a greater concern was the pain in my wrist and the worsening numbness in my fingers. My wrists were bound too tightly, cutting off circulation to my hands. At first I said nothing, not wanting to draw any attention to myself in case that might make things worse.

Some thirty minutes later, I could no longer feel my hands. Worried I might suffer permanent damage, I took a risk and asked, as politely as I could, "Would you please loosen the bindings on my hands? The circulation is cut off and this is not good. It could hurt my hands." I braced myself for a slew of potential reprisals: hitting, kicking, another sandal in the face...maybe being bitten or mauled by the unseen crocodile-thing that seemed to amuse them. They ceased their chatter for a moment, which only intensified the gurgling sound. Then, nothing. No slaps, kicks, or sandals. But also no relief for my numb hands.

Since they didn't seem inclined to punish me, I gave it another try. Maybe they had misunderstood. I spoke a little louder, "My bindings are very tight. I'm afraid they're going to cut off the circulation to my hands. Will you please loosen them?"

This time a menacing voice replied in English, "Do you want to complain about it?"

I shook my head and fell silent.

Over the next half hour, the strange noises subsided, the smoke dissipated, and I began to believe I might actually lose the use of my hands. Then someone stood me up and walked me a few feet to the center of the room, untied my hands, and removed my blindfold. A quick scan of the large hut revealed a hearth-fire in the corner, a couple of machetes and pots, a few bamboo sleeping mats with mosquito nets, but no animals of any kind. What had made that awful sound?

I stood facing an ordinary wood desk, something a 1950s schoolteacher might use. Behind it sat two people. One was a guy in his twenties, wearing a pith helmet with the now familiar Vietnamese gold-star insignia. He held an open notebook and a pen. The other was about 50, with graying hair and a goatee, an olive-drab military shirt with the sleeves rolled up to the elbow, and a messenger-bag slung over his shoulder. He wore a baseball hat tipped at a ninety-degree angle, much like Soupy Sales. I fought the urge to laugh at that image. A pack of cigarettes had been tossed on the table in front of him. Although he was casually dressed, his look of deliberate indifference suggested he was the one running this operation.

To the side of the desk, the fellow who had been riding shotgun in the jeep stood at something approaching military attention. On the floor against the hut's bamboo wall sat a small family: a woman, a couple of young children, and an old man. This was apparently their home.

I was still dressed in nothing but my underwear.

The young man in the pith helmet asked in English, voice firm but calm, pen poised to write, "What is your name?"

"Robert Wideman."

"What is your rank?"

"Lieutenant Junior Grade."

"What is your serial number?"

"689953."

"What is your birthdate?"

"October 17, 1943."

"What city are you from?"

I hesitated. Our eyes met. His gave away nothing. I tried to make mine equally blank. It seemed such an innocent question, and if anyone back home had asked it, I would have told him. What harm could there be in naming my hometown? It wasn't a military secret. But I knew that the code of military conduct only requires prisoners to give name, rank, serial number, and date of birth. If I answered anything beyond those four things, it could turn into a slippery slope. Surely he was trying to pry open just a small corner of the lid, and if I let him in who knew how far he would take it? What's more, thoughts of my hometown were full of emotion; there was no place I would rather have been at that moment. If I gave him that, it would feel like a violation. It would be the beginning of my undoing.

Only a few seconds passed before I answered politely but firmly, "I'd sure like to answer you, but my government will not let me answer that question."

He gave me a stern look, his voice under tight control. "You must confess correctly."

I interpreted that to mean that the sorts of questions I imagined would come next would indeed come next: he would expect me to give up strategic military or political information. For the first time since my capture, I identified a feeling in myself other than homesickness or vague apprehension. Not quite fear, not yet. More like the jitters that precede fear.

A couple of weeks earlier, our air wing intelligence officer had briefed us on prisoner conditions: "We don't know much about what's happening up there, but we believe they're getting good treatment." Then again, an admiral had also said that our raids were stopping more than 98 percent of supplies, and the endless line of trucks I had seen the night before told me that was false. So maybe the good-treatment line was false too.

I braced myself. I had no real way of knowing what these men might do to me. I only knew that I was an American, and therefore their enemy. I still believed in my patriotic duty, and right now that was to say nothing, reveal nothing.

"What city are you from?" the one in the pith helmet repeated.

"I'm sorry, sir"—during our training we had been taught to use the term "sir" if we were ever captured—"I'm not permitted to answer that question."

He slammed the desk, and I tried not to flinch. "You must confess correctly."

I repeated myself.

"You must talk slower," he said. "I cannot understand you."

I could barely understand him either, because of his accent, though I didn't say so. I talked more slowly, just as he asked, but still politely declined to answer. He insisted I "confess." A few times I responded, "What?" indicating that I didn't understand—whether I did or not. This went on for several minutes.

Then he rose to his feet. "If you do not confess correctly, I will have my men garrote you."

I pictured the Russian agent trying to strangle James Bond to death with piano wire in *From Russia With Love*. This is gonna hurt, I thought. Then I saw the two-foot length of thick bamboo rope lying on the table, and for me that was a strand of hope. That's way too thick to garrote me. He probably doesn't know what garrote means. He probably just means he's going to have the guard tie me up, just like I've been tied up for the past hour. No big deal. Stay calm.

For a minute we stared into each other's eyes, waiting. It was not silent. I heard the *whomp, whomp, whomp!* of exploding bombs moving up the hill toward us. The older man with the Soupy Sales hat reached across the desk to a small kerosene lantern with a cover on it and turned down the flame. It occurred to me that the cover was meant to prevent light from diffusing upward during night raids. When had night fallen?

Hearing the engines of American bombers growling nearby gave me a surge of resolve, even though I knew one of their bombs might land right on this hut. I might be obliterated, but Soupy and the interrogator in the pith helmet would go with me. I stood up straight and looked the interrogator in the eye, ready for anything. He stared back, eyes flat with the gravity of the situation. Silently we waited to see how this moment would end. Then the planes left.

He tossed his notebook onto the table, not in a threatening way, more like a teacher frustrated with an underperforming student. "I will give you five minutes to make up your mind." He stood up sulkily and walked out.

The others stayed put and muttered among themselves. The family continued to stare at me, likely wondering what I would do next. I wondered too. What would happen in five minutes if I didn't "confess correctly"? It wasn't long before the interrogator strode back in. This time he did not sit, and he failed to maintain his former measured coolness.

Instead, he paced back and forth, screaming the same thing over and over, leaning toward me each time he drew near, bathing my face in a mist of spit. "You must confess correctly! What city are you from?!"

I repeated, "I'm sorry, I cannot answer that question."

He sat down, lowered his voice, and continued to play the game a while longer. Then he said, "I will give you another five minutes. Last chance. Five minutes." He rose and left the hut again.

Nothing had happened after his initial five-minute warning, so maybe this was another bluff. I didn't think *I* was bluffing, but I honestly wasn't sure. I tried not to think about the garroting, which would either strangle me or slit my throat. Hopefully I would die fast.

This time he was back within a minute. His voice was eerily calm. "If you do not tell me where you are from, I'm going to have my guard take you out and shoot you."

That worried me even more than the idea of being garroted, as it seemed a more ordinary and realistic possibility. I'd seen plenty of rifles pointed at me over the past two days, and this was, after all, a war. I had watched a movie made in 1966 about the Hanoi March, and I had recognized some of the pilots in the movie. Those guys didn't look too good: emaciated, filthy, covered in sores, eyes listless, looking ready to drop dead. So maybe the treatment of prisoners wasn't as great as our superiors had told us. I couldn't help thinking it was unlikely this guy would face much in the way of consequences if he killed me. It was unnerving to say the least, gauging the likelihood of another man's resolve to follow through on a threat to end my life.

I took a gamble on my captor having retained his basic humanity despite the war. I scrunched my face up and did my best to squeeze out a tear. I felt so tense that it wasn't difficult. The tear trickled down the side of my nose, and I resisted the urge to wipe it away, hoping it would convince him I took his threat seriously but had decided to let the guard shoot me anyway. Maybe if he felt certain that the threat of death would not convince me to talk, he would see the futility of killing me and instead keep me around as a potential asset.

I told him for the umpteenth time that my country would not allow me to answer his question. He slammed his hand on the desk so hard that the children along the wall jumped. Then he strode out of the hut again, returned a minute later and said, "Okay, now I'm going to have my guards tie you up."

Two guards stepped up and forced me to kneel in front of the desk. They picked up the bamboo rope and tied it just above each of my elbows in a slipknot. Then they cinched the knots with steady pressure until my elbows were squeezed into nickel-sized knots. The pain was mind-blowing, like a white-hot barbed wire slicing deep into the flesh of my elbows and plinking every nerve along the way until it reached bone. They then used the rope to squeeze both elbows together behind me like a trussed chicken, and pulled.

It was unlike any pain I had ever known before. A fist punch, a football tackle, a knife cut: all were momentary shocks that would hit hard and then recede into an ache or throb. But this, this was an endless zinging, piercing, crushing, squeezing, blasting agony that all but made my very marrow scream in horror.

Blood trickled from beneath the rope where it broke through the flesh, the sort of thing that gives a guy a front row seat to what his execution might look like. I was determined to take it. I didn't want them to think I would easily betray my country. Hell, I didn't want to think it myself. But within the first few minutes the agony felt eternal, as if I'd always been in this much pain and always would be. "What can a prisoner do?" I wondered. Certainly my resistance had its limit. When would I reach it?

After about half an hour wrapped up in that rope, my arms were so purple they appeared blackened, beyond a mere bruise. It resembled gangrene. Occasionally one of the guards would step forward and tighten the rope another notch until I was certain it couldn't tighten any further without breaking my bones.

Between the relentless pain and the fear of losing my arms, I could think of only one course of action: Pass out. If you pass out, the pain will stop. Then maybe they'll leave you alone. It didn't feel like it would take much for me to lose consciousness. My vision clouded, I fell forward, and then *whap!* My forehead hit the corner of the table, and, instead of knocking me out, the shock of the blow jolted me awake. I heard a chorus of gasps from the family that still sat in the room with us. One of the two guards behind me grabbed my hair and yanked me back to a kneeling position.

When the guard stepped away, I tried to faint again: heard the watching family suck in a collective breath, watched them fade into a blur, felt myself fall, and then *whap!* Up I came again amid a great gasp from my small audience, and a high-pitched sob that must have come from the mother, as if they all suffered with me.

After several tries, it dawned on me that I wasn't the first man to attempt the fainting technique, that this entire set-up was designed to take fainting into account. I'll admit that some distant part of my brain saw the genius in their technique, which allowed them to inflict maximum pain while still keeping the victim conscious.

Floating somewhere above me, my tormentor's voice repeated again and again, "Are you ready to confess correctly?"

I said nothing. Then the kid who had ridden shotgun in the jeep appeared above me, his shotgun replaced by a raised Ho Chi Min sandal made out of old tire treads. The black treads smacked my face repeatedly, splitting the air like a whip and likely turning my face several shades of red, but the sting barely registered above the pain in my arms. Oh God, my arms!

"Are you ready to confess correctly?" The interrogator circled around the desk, grabbed my hair, and held my head back so that I was forced to look up at him, upside down and even more foreign to me from this angle. I may as well have been on another planet, except for the bombs dropping in the distance. The ground shook and rumbled, but it was difficult to care. My agony consumed me. Water poured from my eyes in a silent, steady stream, though I was not crying in the usual sense. It was an automatic response to the searing sensation in my arms. All I could think was how black my arms looked, how they might have to be amputated if I ever made it home.

Seeing the direction of my gaze, the interrogator leaned down and whispered in my ear, "If you stay like this much longer, your arms will fall off."

Yes, I thought, they probably will. If I had any hope of dying, I wouldn't have cared about my arms. If I died, this would all be over: the pain, fear, uncertainty, the absolute loneliness of the past two days. But I knew from the look on my captor's face that he would not release me into the relief of death.

All because I did not want to admit I was from Cleveland.

After about forty-five minutes, I felt a sort of convulsion rising from deep inside me. I cut loose and screamed as loudly as possible, as if some animal part of me hoped that howl could take the pressure off my bloody, blackening arms and the rest of my aching body. And the word I screamed at the top of my lungs was, "Never!" I repeated that word at least three times, louder each time, "Never! Never!! Never!!! I'll never tell you!" I was in the final throes of an impassioned fantasy: I was the American fighting man who would die before he gave in to the enemy.

Oddly, that very moment of refusal was also my moment of surrender, my breaking point. Not five minutes later, I said, "If you untie me, I'll tell you anything you want to know."

The guards immediately untied me. The pain dissipated almost as quickly as it had come. I looked down and watched the blood pump back into my arms. Within ten minutes they were a faint pink and felt more or less normal again, though there were faint red ligature marks and dots of blood circling my upper arms just above the elbows. Looking at my almost ordinary looking flesh, I thought, Good God, you gave up so easily.

"What city are you from?" the interrogator asked.

"Cleveland, Ohio." How simple, how easy, all that pain seemingly for nothing. But I told myself it had been necessary to prove, if only to myself, that I had meant to be strong, that I had wanted to be loyal, that I was still a man, however weak.

"What kind of an airplane were you flying?"

So, I was right. My willingness to answer one unauthorized question had been intended to lead to another. But I was done. I had put up my resistance. It was over. "A-4."

"What squadron are you from?"

"VA-93."

When I told him my skipper's name, he terminated the interrogation. Two guards half-led, half-carried me to a cubby in the corner of the hut. The cubbyhole was only about three feet by six feet, separated from the rest of the hut by a hanging bamboo mat. One side faced onto the larger room, so I had no privacy. At the far end of the cubbyhole was a thick pole that looked like a telephone pole driven into the ground, probably a support for the hut. They laid me down on a board on the dirt floor, apparently to keep me off the damp earth, though my legs dangled off the end of the board. They then fixed a two-foot chain around one of my ankles and wrapped the other end around the pole. I had less room to move than a dog tethered in a yard. Then they tied my wrists and hogtied me to the board. It was uncomfortable but bearable, nothing like the pain I'd just endured.

The interrogator appeared above me like a giant in the lamplight. "Okay, Robert"—he pronounced my name Row-bert, "Now you may sleep. I will talk to you in the morning."

By that, I understood that I would be killed first thing in the morning. I had been told that if I ever told my captors what they wanted to know they would have no reason to keep me alive. I had given in. This must be my last night on earth. I barely had energy left for regret or fear. At least the pain was over. The idea that I would be shot come daybreak was the first certainty I had felt since my plane had started to dive, and there was a strange liberation in that. I fell fast asleep, no longer thinking about what was coming next.

5

AMERICAN DREAMS

I could not bring myself to consider the bloody cinching of my arms as torture. But that night I knew that the ropes tying me to the board and the chains shackling me to the pole were not a matter of opinion. They were solid and undeniable. I felt the need to urinate, but was so exhausted I couldn't wake. Instead, I dreamed I needed to sit up and pull on a big iron ring attached to the post in front of me. I struggled to reach it. What a relief to finally pull on it and release my bladder! I woke, looked down, and saw that I really was sitting up, my crotch soaked in warm urine. The smell was putrid.

The jangle of my chains, the sound of peeing, or my small groan of disgust must have alerted the guard sitting outside. He stepped into the room, which was dimly lit by a lantern. He stared from the piss spreading in the dirt to my soaked underwear and shook his head. He pointed at the mess, rubbed one of his index fingers against the other, and scolded me, "Tsk, tsk, tsk." It was bad enough I had no control over what my captors did to me; it only made matters worse to lose control over myself.

Moments later I saw what looked like a big toad sitting on the ground staring at me. It was unlike any toad I'd ever seen, with huge teeth like a gopher. Was this the gurgling creature I had heard earlier? Had it come back to bite me? Might it give me some fatal tropical disease? I called for the guard, who returned with a sigh of annoyance.

"There's an animal in here with sharp teeth!" I turned to point it out, but it was gone.

The guard rolled his eyes as if to suggest I was losing it. Maybe he was right. Still, I was never fully convinced that the toothy creature was just a dream.

The sun began to rise, and a rooster's insistent crowing dragged me from sleep. The sounds of a bubbling stream, waking birds, and leaves rustling in the morning breeze breathed hope into me again. Then the smell and dampness of my urine-soaked underwear brought me back to reality. Voices muttered outside. A shuffling in the hut announced the return of my interrogators.

The Vietnamese Soupy Sales stepped into my cubbyhole. "So, now you are my preeeees-ner!" His words jolted me like caffeine. He had not spoken English the night before, had not spoken much at all. Perhaps it was as I had thought initially: he was the one in charge. My submissiveness of the previous night came back to me. Peeing myself was nothing compared to that. He untied me, unlocked my chains, and helped me to my feet—still a laborious process. He led me to the outer room to stand in front of the desk again. Then he sat behind it.

"How is your health?" Soupy's politeness was almost scarier than the other man's shouting from the night before.

I raised an eyebrow. "As good as can be expected."

"Do you have enough to eat?"

"Yes."

"How did you sleep?"

"I was a little bit cold."

His tone became jocular. "Oh no! You cannot be cold because you are from Cleveland. It is much warmer here."

"Yes, I know, but in Cleveland I had clothes, and here I don't."

"But in Cleveland there is snow. It is much colder than here. You should not have problem."

I did not argue. My life was in his hands.

"Are you married?" So, again we would start with innocent-sounding questions, but where would it go from here?

"Yes."

"Do you have any children?"

"No." He sat up straighter, proud I thought.

"I have one child." I would come to learn that every interrogator always had one more child than the prisoner did.

I said nothing. I had already capitulated. The best I could do now was to refrain from offering any more information than necessary.

The questions continued in a non-military vein. Did I like sports? Did I like music? Did I like art? I wondered whether he was merely curious or trying to throw me off guard. He told me he knew the famous American physicist Linus Pauling. He said that he used to be a math teacher for schoolchildren, that he loved his old job, but that the war had forced him to become an interrogator. I wanted to believe him, to think of him as a reluctant warrior who would rather be teaching. It was hard to resist telling him about myself, my wife, my dreams of becoming an airline pilot. It was tempting to think of this as a sort of cultural exchange program. It was a clever manipulation.

When Soupy began asking military questions, I thought I detected the kindly old schoolteacher he once was, patiently drilling students for a test. "What were your dive angles?"

"My government won't let me answer that question."

"What were your release altitudes?"

"My government won't let me answer that question."

"Do you remember what happened to you last night?" he chided.

"Yes."

"So, what were your dive angles?"

I tried to keep my answers general. "Between five degrees and ninety degrees."

"What were your release altitudes?"

"Between one hundred and ten thousand feet, depending on my flight leader's preferences."

"What was your last target?"

"You already know that because your people shot me down over my last target."

He went on to ask me the range of various American weapons I had used and what sorts of electronics I relied on for night flying. At one point, he brought out my survival pack and asked me the purpose of every item.

"What is this?"

"Aspirin."

"What does it do?"

"To help with headaches, fever, or pain."

I figured he knew what it was, that his purpose was to keep me talking just to prove he could, that this was how he had been taught to establish power over prisoners. I did my best to undermine that power.

"What is this?" He held up a signal mirror.

"Well, it looks like a mirror to me." I tried to sound cooperative, not a trace of sarcasm.

"What do you use it for?"

"I don't know."

He turned it over. "We have some directions on the back." He showed them to me.

I leaned forward, glanced at the instructions. "Oh yeah. How 'bout that?"

"What does it mean?"

"I don't know. I'd have to take a look at it. I've never used one before."

He handed it to me, and I took my sweet time reading and attempting to execute the instructions. They were simple enough: look through the hole in the mirror's crosshairs, find the sun, line it up, and reflect sunlight off the mirror. Of course, I knew I couldn't find the sun's position from inside the hut, but I spent several minutes pretending to try.

He snatched it from me, shaking his head with an expression that was easy to translate: You dummy. How can an aviator not know how to use a signal mirror?

I bit my lip to hold back nervous laughter.

He returned to non-military subjects. "Do you own a house?"

"I have an apartment."

"I have a house. Do you have a TV?"

"Yes."

"Color TV?"

"Yes."

His eyes had a faraway look. "Someday I will like to have a TV. Do you drive a car?"

"Yes."

The interrogation lasted a couple of hours. Then Soupy left and one of the guards tied me up again. This guard was more easygoing and left the knots loose, smiling at me and patting my sore arms. The guard and I remained in the large room as the hut reverted back to a home and gathering place where several generations of family members came and went.

During the midday siesta, as I dozed in my cubbyhole, several young men gathered in a circle on the floor of the main room, and that's when I discovered the source of the crocodile sound from the night before. It was a foot-long bamboo water pipe. It looked like a bong, but when the easygoing guard took a pouch out of his pocket, it did not contain marijuana. Instead, he pulled out a pinch of loose tobacco and stuffed it into the pipe's spout. He then took a sliver of bamboo to the hearth-fire that almost constantly burned in a corner of the hut, walked back to the circle, and used the burning bamboo to light the tobacco while holding his mouth to the pipe's wide end and inhaling. That's when I heard it: the gurgling sound was bubbling water. The guard inhaled for a long time before he took his mouth off the pipe, held his breath for a few seconds, then let it out suddenly, much as a marijuana smoker would. He passed the pipe on to the next man in the circle.

The bamboo water pipe made its way back to the guard, who walked over to me and held it out with a grin. It had been a while since my last cigarette, and I was curious about what it might be like to smoke a water pipe, so I let him show me how. I peered into the pipe and the liquid inside was thick and black and stank of tar, which made me nauseous, but not enough to end my curiosity. The guard lit the pipe while I inhaled, then he mimed that I should hold in the smoke. It seared my throat and expanded my lungs until I coughed a huge cloud. I hacked so hard the guard whacked me on the back while the men in the nearby circle laughed. I felt dizzy, ten times dizzier than the first time I had smoked cigarettes, as if I had just smoked a whole pack of Camels in a single puff. But it was not pot; I did not feel high.

Years later, I would learn that this way of smoking is an old North Vietnamese tradition. The small bong is called a *dieu cay*, or farmer's pipe, and the tobacco is a locally grown crop called *thuoc lay*. I also later learned that sometimes farmers cut their tobacco with opium, but if my guard did such a thing I never noticed it. My aches and pains remained. I did not float away from my troubles. My situation did not suddenly look rosy or amusing. Still, I did like the taste and feel of the tobacco, which did enhance the typical relaxing effects of a cigarette, and gave me something to do besides dwell on the idea that I was an animal in a trap.

Since the easygoing guard had been generous with his tobacco, I risked asking him if someone would wash my skivvies, which were caked with mud, sweat, and urine. I could smell myself and it wasn't pretty. Wonder of wonders: he gave me a pair of green pants and a faded shirt with a high Mandarin collar. I put them on and he took my skivvies away. I breathed in the freshness of the clean clothes, a smell of the outdoors, of freedom. I never knew I could be so happy to feel almost clean. If only I could bathe, but I didn't dare push my luck. That night I did not feel chilled and I slept without waking. I felt almost human again.

The following week I fell into the routine of my captors and learned the greatest enemy of a POW: boredom.

After the rooster crowed each morning, a guard took me outside to a pigsty to hang my genitals or my bum over the little bamboo fence and urinate or defecate. The sty was full of fermented food-waste and manure—both porcine and human—and the stench was so intense I almost vomited the first time. I soon grew accustomed to it, and one time I even peed on the pig's head when he got in my way.

For two hours each morning, I returned to the big room for interrogations, sometimes with Soupy, sometimes with a different interrogator. The alternate interrogator seemed as bored with being a captor as I was with being a captive. He would spend an hour and a half out of our two hours asking how many TVs I had, which razor blades I used, what kind of food I ate. However humble my answers, he nodded as if I confirmed his thoughts, saying something like, "Someday I will like to go to America." He did not sound as if he begrudged Americans their luxuries, but as if he were truly interested in all things American. Still, he could not resist an opportunity to one-up me: "I have five children!" Five more than I had, so he was rich in something.

Sooner or later they would always return to military questions, questions about the range of my missiles or capabilities of my aircraft. Sometimes I generalized. Sometimes I flat-out lied.

"What's the range of the Bullpup missile?" Soupy once asked.

"About a mile." It was actually about three or four miles.

Soupy, who clearly had no knowledge of aircraft, seemed willing to accept my answer.

But a young man wearing civilian clothes had joined us. He tightened his jaw, shook his head, and muttered in English, "Too short. Too short."

My eyes darted from him to Soupy, waiting for one of them to threaten me with the rope again. But they let it go.

At some point, each interrogator would light a cigarette for me. Soon that seemed barely sufficient to tide me over until my next toke from the tobacco bong.

In the middle of each day, the village fell into its usual siesta, during which I too dozed. One day as I woke, the easygoing guard strode in, pulled a little bottle out of his pocket, and gestured to me to hold out my hand, which I did.

He shook about half a dozen yellow pills into my hand and said, "Vitamin-mineral."

I stared at them, hesitating. How did I know they weren't poisonous?

Seeing my trepidation, he poured a few into his own hand and tossed them in his mouth. His smile and nod seemed to say, See, it's safe. It's good for you!

Then someone approached the hut, and the easygoing guard motioned for me to hurry up and take the pills, as if he feared we would be caught holding the bag. I tossed the pills in my mouth and swallowed, almost choking with no water to wash them down. He gave me a conspiratorial look of relief as my interrogator entered. I never knew what was in those pills. Maybe vitamins just as he said, maybe something else. Either way, I did not feel the slightest effect.

In the afternoons, it was time for round two of the day's interrogations: another two hours of question-and-evasion, cat-and-mouse, empty cultural exchange and military intelligence questions, all interspersed with threats of physical punishment.

After a typical afternoon interrogation, it was time for another tasteless meal of rice and greens. Then the young people would often light up another bowl of tobacco. And I do mean young people—sometimes children as young as seven or eight would puff on it. I was always a little excited to see the bamboo pipe come out, eager for the nicotine fix, and eager for something, anything, to do, if only for a few brief moments. It seemed there was no danger my hosts would cut me off. They got a kick out of my ineptitude with the pipe and how much I coughed when I smoked.

At sunset, a guard took me out to relieve myself. After that, he returned me to the board in my cell, blindfolded me, tied my hands, and chained my legs. Then I did my best to sleep through unrelenting hours of chilly darkness. If I couldn't sleep, I tried to entertain myself by recalling old songs, old stories, old shows. One question continued to elude me: Who was *The Shadow*?

One night I had to pee after dark. The easygoing guard was not on duty. It was someone else. He secured my blindfold, which tended to slip off in the night. Then he tied my hands behind my back, grabbed one of my arms, and guided me outside. I took several steps before one of my feet stepped through air and I dropped about four feet into a trench. I landed squarely on my crotch, and the pain that zinged up my spine was so intense that for a moment it felt as if my already aching backbone was broken.

I heard the guard chuckling as he held out a hand and waited for me to find it in the dark so he could pull me out of the hole. I felt certain he had misguided me on purpose, but I said nothing. What good would it do?

Most of the guards seemed intent on reminding me of my place, tightening my ropes until they hurt, tightening my blindfold until I saw spots. But whenever the easygoing guard was on duty, he undid my blindfold and untied my hands, though he left the blindfold and ropes hanging loosely around me so he could motion to me to quickly put everything back on whenever he heard a superior coming.

One night after everyone fell asleep, he brought his girlfriend in and introduced me. She and I didn't say anything special, just "Hi!" But with that small gesture he showed me so much more humanity than anyone else. That emboldened me. One night, he brought in some paper for me to draw on. I drew a map of Vietnam, which seemed to amuse him. He nodded and smiled over how well I seemed to know his country. Then I drew a giant arrow leading out of the country. I pointed at the arrow, then at him and me, and used two fingers to mime walking.

Leave the country...you and me...together?

He stopped smiling, sat back on his heels, and shook his head firmly.

I pressed my point for a moment. "Help me? America?" I rubbed my fingers together to suggest cash. "I'll pay."

He shook his head harder, gave me a regretful look, crumpled the paper and walked away with it.

I wasn't especially disappointed. I had known he would be unlikely to do it. Surely he would be killed for treason if he were caught helping me escape. Still, I felt satisfied with myself for having tried. It seemed important not to give up, not to count myself as lost. If I accepted my situation, it would feel like the enemy had already won.

A few days into my imprisonment, the guards replaced my chains with rope. With that, I began planning an escape. During the siesta, the guards sat outside and left nobody inside to watch me. I looked out the door at mountains that rose a tantalizing five miles to the west. I knew that Laos lay another ten miles into those mountains in an area much less populated than the delta, and I knew that American reconnaissance aircraft patrolled a safe area in the Laotian corridor. If I could get there without being caught, if I could leave a sign, and if a pilot saw it, he might pick me up.

I scouted my tiny room for supplies. I squirreled away an extra shirt, a blanket, and a small pair of children's rubber-tire sandals that would at least protect my feet a little. I hid my trove in a corner under a pile of unused mats or other household odds and ends. I spent most of two siestas searching for matches and a container to boil water, but those two items eluded me.

One day the guards outside the door were gone. Nobody was sleeping in the hut, walking on the path, or working the fields. It seemed everyone was asleep but the whirring insects. The soft green fields of the countryside floated in golden sunshine. The dark green mountains rose beyond like a gateway home. I could almost hear the warm air whisper, "Escape. This may be your only chance. Nobody's watching. Run for it! Freedom lies just beyond those mountains."

I still had no way to carry or boil water, but such practical considerations seemed meaningless at that moment. I only had to want freedom badly enough. The easygoing guard had left my bindings loose as usual. There was nothing to stop me. I rose to my feet and charged for the door. That is to say: I took three steps, each one slow, unsteady, infinitesimal. Each step took me mere inches toward the door. I walked slower than the eighty-year-old grandfather who slept in this hut every night. Every muscle, ligament, and tendon felt like a rubber band stretched thin and ready to break. I stared at the dark shadows of the mountains.

I tried to will my body forward. I told myself what I had learned in training: if you're going to escape, you must do it in the first few days; once you get to an organized camp, it's all over; after that, your captors will always have the upper hand. Another voice in my head responded that even if I had all my strength, someone would spot me crossing the rice paddies and catch me within thirty minutes. Even if they didn't, I might contract dysentery from drinking stream water and shit myself to death before anybody found me. In the end, the decision wasn't up to me: I could barely move.

I never made it to the door.

The interrogators and I developed a familiarity with one another. My enemies were my only companions and I knew this was dangerous. I did my best to keep my guard up, but it was exhausting and my mind often wandered. I was bound to slip sooner or later and say something I would regret. Sometimes they interrogated me into the dark hours while bombs fell nearby.

One night the bombs landed so close that the percussion of ground and air vibrated me from the inside out: *crack-thud, crack-thud, crack-thud.* Soupy turned down the lantern, hand shaking, and stood still, listening. Sweat dripped down his temples as we waited. It sounded like a giant was thundering toward us, but I felt less fear at those moments than any other.

He stared at me. "It does not scare you?"

"No. I know they're not going to bomb here. There's no target around here worth bombing. If I was in Hanoi maybe, but not here."

My calm only agitated him more. "Why have you come to bomb our country?"

"Because my country sent me here."

"You bomb the roads at night, and it is a Vietnamese custom to walk on the roads. The people walk on the roads, and you come and you bomb them and you kill many civilians."

That idea made me feel bad, but I refused to let him see that. It was a mental trap. This was war. The communists were the bad guys, bullying the South Vietnamese just as he was accusing Americans of bullying the North Vietnamese. I wouldn't play his game. I said nothing.

"Why do you do this?" he insisted. He would not let it go.

Knowing he would not let up, I gave the most neutral answer I could think of. "You're sending supplies to communists in the South. Those communists are trying to take over South Vietnam, and they're killing Americans. I'm bombing your country to stop those supplies."

He slapped the desk. "No, that is not true! Americans are the aggressors. This is not your country. We did not ask you to come. You are committing aggression."

"I'm here to bomb the communists' supplies. You're supplying the Viet Cong."

He seemed exasperated with my stupidity. "I *am* the Viet Cong." I did not know then that *Viet Cong* can simply translate as *Vietnamese soldier.*

"Oh. Then that's why I'm bombing you."

"You must not use that line of logic anymore. It is wrong. I tell you it is wrong. If you use it again, you will be punished severely!"

I didn't answer anymore, but his voice stuck with me through the night: "You are the aggressors!" I countered with my own inner voice: Hey man, if your government wages war, you've gotta expect to take some losses. But really, I didn't know what to think.

Did civilians deserve to be punished for what their government did? Soupy didn't want to be there, he wanted to teach. Maybe the civilians didn't fully support their government's war. Should ordinary people crossing bridges and roads and waterways, people going about their day-to-day business, expect to die because they lived in a war zone? On the other hand, what the hell were we supposed to do, let the NVA slaughter all the South Vietnamese who got in their way? Let communism inch across the world until it swallowed everything?

I didn't think about it for long. A guard brought my skivvies back, clean and white, and took the borrowed shirt and pants away. I was cold again that night. At that point, I didn't care whose fault the war was. All I wanted to do was drive the aching cold from my body so I could sleep. I wrapped my arms around myself, but it wasn't enough.

I don't know if he got tired of debating me, or if my time there was simply up, but I didn't see Soupy Sales again. On May 15, after supper, a guard walked in and said, "Okay, move to new camp." He walked me through the village to an old beat-up pickup with wood sides.

A man in a pith helmet stood behind the truck, grinning, holding out his hand. I didn't recognize him at first because of the helmet. I was pleased and relieved to see it was the easygoing guard who taught me how to smoke a water pipe, loosened my ropes, and gave me "vitamins." I shook his hand and gave him a grateful smile, though I didn't dare say a word. He smiled wider, as if our exchange made him happy, and he shared a cigarette. I climbed into the truck bed, and he walked away.

6

ORANGE SKIES

In the dark, it took me a moment to notice that an American wearing nothing but a t-shirt, boxer shorts, and a blindfold was sitting in the truck bed with me. I was still stripped to my underwear too, but not blindfolded. The guards chained our legs together. They had taken my chains away three nights before, and I now realized they must have needed them for this man, whom they had kept elsewhere in the village. Since he was newer than I was and, judging from his appearance, less injured, maybe they assumed he posed a greater flight risk.

He was the first American I had seen up close since I left the carrier on my last bombing run, and since he was blindfolded I took the opportunity to study him. Ever since my crash, I had been anxious to see another American. Now that the moment had arrived, it was strangely deflating. Looking at him was like looking in a mirror for the first time since my capture: he had several days' growth of beard and a sunburn, his skin was grimy and sweaty with speckles of blood, and like me he reeked of pigsty. I wondered if my face had the same emotionless look. He had a crew cut typical of a pilot, and most Americans captured in the North were pilots, so I assumed he was one. But he did not look familiar; he was not from my carrier. Just another poor stiff with rotten luck. Looking at him made it clear to me: You're a prisoner, Bob. Fuck.

If I harbored any ideas about improving our acquaintance, the half-dozen armed guards who sat in the truck bed with us nipped that in the bud. One of the guards blindfolded me, and I could no longer see my comrade. The guard then untied my wrists so I could hold on during the ride.

I tried to offer a friendly tip to my fellow American, figuring he was at least a few days newer to the prison transport business than me. "If you lie down and rest on your elbows and float with the bumps, you won't get as banged—"

One of the guards struck me on the shins with a rifle butt. "No talk!"

I grunted involuntarily at the pain but otherwise fell silent for the next several hours.

It was a rough ride. Even with my hold-and-float technique, it was difficult to hold on and stay upright in that rusty old rattletrap as it bounced over rutted, muddy, unlit roads. Although the driver never reached ten miles an hour, my spine took a beating.

Every hour and a half the droning of jet engines approached. The driver stopped and cut the headlights, and we quietly listened to the thudding of the nightly American bombing raids. Sometimes the truck shook with the impact. The first few times it wasn't alarming, just another routine of war. If the bombs hit anything but earth, foliage, or crops, I didn't hear anything to indicate significant destruction or loss of life: no crackling fires, secondary explosions, or screams of pain. I imagined pilots I knew flying those runs, blindly missing target after target, gouging empty fields with pointless craters.

After about four hours of that, it began to rain. Hard.

The guards ducked under a big green canvas tarp lying in the back of the truck, but left the other prisoner and me exposed to the torrential downpour. I didn't blame them. It seemed to me there wasn't enough room under the tarp for everyone, we were not part of their team, and we both stank like hell. Had it been daytime I might have welcomed the rain as the closest thing to a shower I'd had in nearly ten days. As it was, the rain chilled me to the bone, and once it started it did not let up for hours. It rained and rained and rained. The more it rained, the colder I felt, and soon I was shaking more violently from the cold than from the rough ride.

After three hours, I felt so frozen my survival instincts took over. Without any thought about how the other prisoner might react, I scooted as close to him as I could, pressed against his side, and tucked my toes first under his armpits and then in his crotch, knowing these were the warmest places of the human body. He was not shivering like me. In any case, he allowed the awkward huddle without shifting away or giving the slightest indication of annoyance. Though I was grateful for the warmth, the human contact offered almost no comfort. I felt as lonely as I had before, reduced to stealing a stranger's body heat, which did not stop my shaking.

Soon the truck got stuck in the mud, its tires spinning in place. All seven guards got out to push, pull, and rock the vehicle. That effort went on for a couple of hours to no avail. Even though they weren't using the tarp at that point, nobody offered to cover us, and I knew better than to ask, what with my shin still throbbing from the earlier strike of the rifle butt. I had spent winters in Cleveland, Ohio and Upstate New York, places known for bitter cold, but this was the coldest night of my life. I became convinced I was going to die, not from the bombs raining around us every hour and a half, but from hypothermia or pneumonia. It seemed a pointless way to leave this world.

Over the next four hours, the guards freed the truck, the rain stopped, and we continued on. The sun was just coming up when the truck stopped, not in Hanoi but yet another village. The guards unshackled us, took off our blindfolds, and gestured for us to jump from the tailgate. The other prisoner jumped first. When I looked down, I knew I couldn't make the jump. It was four feet to the ground and my leg muscles felt rubbery after my ejection from the plane, the long nights of walking, and my long soak in the cold rain. I sat on the edge of the truck bed and held out a hand to indicate I needed help. The only response of the guard below me was to shout in Vietnamese and gesture impatiently at me to hurry up.

I leapt off the back of the truck. For a split second, my feet touched the ground, but my legs no longer felt like part of my body. They buckled under me, landing me on my tailbone so hard I could hear the loud *whap!* as I hit the compacted dirt. Embarrassed at my weakness, I attempted to get up without help. Yet again, it took too long, and a couple of guards hauled me back onto my feet.

They led me, staggering, about a hundred yards into a large gray concrete building with a thatched roof, about the size of a gym. The inside was partitioned into two large rooms. They took me to the far end of one room and the other prisoner to the far end of the other. Then they laid out bamboo mats for us on the dirt floor. Glassless windows illuminated the dim space with squares of sunlight that suggested warmth without giving it.

I huddled into a fetal position, shivering, my teeth chattering like the wind-up false teeth that novelty companies sell, knees knocking like castanets. I tried to hold my knees still, to squeeze them together with my arms and put my hands between them, but nothing would stop them from knocking. The guards admonished us again not to talk, so my partner said nothing, but he later told me that my knees had clacked together so violently he could hear them all the way across the building. A guard brought me a cigarette, and I could barely get it into my mouth with my trembling hands.

During the long drive, one of the guards must have taken pity on me and given me a t-shirt, though I couldn't remember how I ended up with it. The t-shirt was now soaked. I figured it would do more harm than good at this point, contributing further to my hypothermia. With difficulty, I pulled it off and held it out to the guards to take away. That seemed to help. About a half hour later I finally stopped shaking.

They brought us breakfast, some sort of rice with peanut sauce, which to my American tongue tasted like rice and peanut butter. Not a combo I would have chosen under normal circumstances, but it had so much more flavor than anything I had yet tasted that it might as well have been a gourmet meal. I must have wolfed it down, because when the guard took my bowl he raised an eyebrow and held the bowl up to me like a question: Would you like another? "Yeah." I nodded so hard it was almost as if I were shaking again. By this time, two whole bowls of food felt like a luxury. I wondered how much weight I would lose before my ordeal was over. I was resigned to the idea that I might be a prisoner for years to come.

After our meal, they brought us the usual hot bland yellow water that passed for tea. Then I lay down again, staring at the other prisoner through the door between the rooms, wondering if I would ever be allowed to talk to him, or to any other American for that matter. Meanwhile, the other prisoner stared back at me, his eyes reflecting similar questions.

As we stared at each other, I entertained hopeful visions of being tossed into a large prison compound like I'd seen in World War Two movies, sharing hardship and jokes with fifty, a hundred, or a thousand other Americans. The best of times, the worst of times. I pictured wearing regular clothes again, covering myself with a warm blanket at night, eating something more than rice and weeds, but mostly talking to other people who knew my language, who came from my country, who were not there merely to torment me and ask me questions. People who did not envy me for owning a television. I imagined making friends, plotting together to sneak extra food, maybe even planning an escape, or at the very least talking about home. The vision was almost exciting. For now, there was only this one man on the other side of the room who might as well have been on the other side of Vietnam. Both in the same boat, but still alone.

The guards must have seen us staring at each other and decided that was too much communication. One of them slammed me in the shins again with the butt of a rifle. Another grabbed some boxes that were propped against the wall and set them between us, obscuring our view of one another.

It must have been about noon by then, siesta time, so I tried to rest. Sleep eluded me. When I closed my eyes, I saw fuzzy versions of my wife's face.

Patricia and I had met in Cleveland at the Lakefront Airport. She was a ticket agent, but to me she looked like a movie star. The moment I saw her, I knew I had a mission: to get her out from behind that counter. I asked her if she could take a break and let me buy her a cup of coffee. I could hardly believe she said yes. As we walked to the coffee shop, I made a few attempts at comedy and her vivacious laugh lightened my heart, a sound so complete and genuine. We must have talked for an hour.

I kept asking, "Are you sure your boss won't be mad?"

She kept replying, "Don't worry about it."

After we drank all the coffee we could hold, she gave me her number.

Now I lay on a hard floor, trying to remember her number, trying to imagine her infectious laugh, trying to bring her face into focus, as if the memory of her might warm me again, brace me against the cold of this country that I used to believe was constantly hot.

I gave up trying to picture Patricia's face and instead studied a group of actual pictures hanging on the wall, including a trio of large portraits, each about three by four feet, featuring the great heroes of communism: a black-and-white photo of the Father of the Russian Revolution, Vladimir Lenin; a color photo of China's Chairman Mao Tse Tung; and a black-and-white photo of North Vietnam's leader, Ho Chi Minh. I had seen images of Ho Chi Minh before, but this one took me aback. In all the photos I had seen in American newspapers, he had appeared serious, dark, and sinister. In this portrait, he looked like somebody's favorite uncle: wise, kindly, and good-humored.

The greatest surprise was the much smaller picture next to the door, approximately eight by eleven inches. In it, a Vietnamese man in a loincloth hung from a cross, a picture of Jesus Christ's crucifixion. I was stunned! Throughout my life, teachers, religious leaders, political leaders, and military superiors had all made clear to me that communists did not believe in God, were not allowed to believe in God. What's more, I was used to a white, round-eyed Jesus. My belief system suggested that Jesus was not Vietnamese, but the image gave me pause. Jesus probably wasn't white either, was He? It was the first time I had ever considered the question.

My mother was a devout Catholic, and I had gone to Catholic school for the first nine years of my life. One thing I remembered from those years was a huge map of the world on the front wall of the classroom. About two-thirds of that map was painted red, with a communist hammer and sickle in the middle of it. I don't remember exactly what the nuns said about all those communist countries, only that they made it clear that communism was bad.

One Sunday when I was 14, my friend David and I went to confession and took communion. David came to my house that evening and told my dad, "You should be proud of your son because he went to confession!"

"That's nice, David," my father said, smiling.

But when David left, Dad exploded. He made it clear that Catholics were brainwashed and the Pope was the chief brainwasher. Then he said, "Never, ever, marry a Catholic woman." This was upsetting, to say the least. Did this mean he didn't love my mother, a devout Catholic woman? I couldn't comprehend how a man wouldn't love the woman he had married. But Dad's word was law, and he laid it down that night. "For the next year, you can go to any church in town, except the Catholic Church! I'll even take you if you want—just not there. After the year is up, you can go back if you still wish to go." I didn't know what to do. I was a Catholic; my mother was a Catholic. How could I just change my beliefs like flipping a switch? I felt that if I went to any other church I would be betraying both God and my mother.

I didn't go to any church that year. After the year was up, I didn't go back to the Catholic Church either, afraid of what my dad might say since I hadn't tried any other religions. My father had shaken the faith I once had that Catholicism was the one true religion. I realized later that my father had simply wanted me to make an informed choice. By the time I joined the Navy, I had no religion, no beliefs, no God.

So it was that during my journey across Vietnam into the unknown, I had no faith to buoy my heart and no God to comfort me. I couldn't remember how to pray, and I was unsure anyone would be listening if I did. What had I given up by letting God go all those years ago? I wasn't sure, but I sensed it must be important if even the communists had given Him a corner of this room. I stared at the Asian Jesus for a long time. Seeing the little goateed Savior unsettled me for years.

The locals didn't even glance at any of the pictures, ignoring them as just part of the background.

After the siesta, the guards took the other prisoner and me outside so another batch of locals could stare at us, maybe a few dozen people. I was used to it by now, and apparently so was the other guy. They stared at us silently. We tried not to stare too long at any of them and avoided eye contact with each other.

The guards then shuffled the two of us back inside for a dull meal, a cigarette, and another toke on a tobacco bong. At sunset, they loaded us into another truck, chained us together, no blindfolds this time, and off we went.

The evening started out clear, but soon clouds began to gather. I prayed silently, Man, please don't rain. Just keep the sky clear. I'm not sure who I thought I was talking to: God? Nature? The sky? I had no idea who was in charge, but it clearly wasn't me.

Maybe somebody heard my prayer. It didn't rain that night. Even the bombs kept their distance. The only maddening thing was the plodding ten-mile-an-hour pace, a slow road to nowhere. We passed supply convoys heading south, forcing us to pull over and wait. We stopped at bombed-out bridges, waiting with a dozen trucks or more to cross via pontoon bridge or other makeshift construction.

At one bridge, we began to pull forward and the POW who was chained to me leaned away from the truck's wooden side, maybe to avoid having his back jostled as we bumped over muddy terrain. Just then a truck headed the other way sideswiped us so hard it smashed in several boards, splintering them inward more than a foot, right at the spot where my silent partner's head had been. Our eyes met just long enough to read a shared thought: if he hadn't leaned forward, that would have been his head and he'd likely be dead. Our driver stopped and flung open his door, as did the driver of the other truck. They faced off, shouting and pointing at each other in the middle of the muddy road. They looked like a couple of cartoon soldiers in a Beetle Bailey comic strip, so furious and animated.

Come morning we stopped in another village to rest and eat, to endure its villagers' stares and its children's hurled rocks. Each hour of this trip began to feel like a recurring dream.

Until the third night.

On night three, we resumed our usual places in back of the truck: six guards, the other prisoner, and me. Within minutes, I began to worry. Though I was blindfolded again, my natural sense of direction suggested we were headed west toward the border, which meant we would soon cross into Laos. Why would we do that? There was no point asking the other POW his opinion or expressing my concern, which would only earn me another blow from a rifle butt.

With nobody to talk to, and no way to search the guards' faces for clues, my mind latched onto Laos and wouldn't let go. Maybe they're taking me to Laos so they can say they found me there and label me a spy. That would give them an excuse to do whatever they want to me. I might be in for some serious torture. I might even be executed.

Such was my frame of mind when the darkness deepened and the nightly American bombing raids resumed. We heard the first jet-engine approach, and the driver stopped the truck and cut the lights as usual. The sound of bombs, missiles, and small-arms fire seemed almost routine. Several minutes later we drove on. An hour and a half later, we heard a second flight approach. The driver halted again, then continued on. Then again a third time.

Sometime in the wee hours, a loud rumbling announced a fourth flight was on its way. This time the truck did not stop.

At first, I was only mildly alarmed. The rumbling increased to a roar. Surely the driver would soon cut the engine and the lights the way he had done up to now. Instead, we rolled on. Something was wrong. The screaming of jet engines vibrated through me, as if we were driving through a vortex of pure sound. I could feel the other prisoner's muscles tense next to me, could feel our chains shift. The guards fidgeted around me. I tipped my head back and peered out from under my blindfold, glimpsed arms gesturing in agitation and legs poised for flight. I heard nervous voices ramping up, up, up with the sound of the jets. One young man beat on the rear window of the truck cab and shouted, another banged his rifle against the window, then another. The truck rumbled on, its faint illumination still broadcasting our location into the night sky. Soon all six guards were slamming the window with palms, fists, and weapons, screaming at the top of their lungs. I didn't understand the words, but I imagine they were something like: "Jets! Bombs! Stop! Stop! The lights! The lights!" They were clearly scared shitless, and the panic was contagious. I wanted to add my voice to their pleas but feared that would make matters worse.

Between the screaming guards beating on the window and the low-flying jets, I found it hard to believe the driver couldn't hear the considerable racket. Perhaps the two men in the cab were talking loudly, and that combined with the closed cab and truck engine were masking the other sounds. Finally something caught the driver's attention and the truck stopped. By then the bombardment was in full swing, closer than ever before, engulfing us in a barrage of sound, each bomb and round of return-fire indistinguishable from the last. Someone dropped the tailgate, the cab doors flew open, and our guards scrambled out of the truck. It looked as if we were all going to run for it, thank God. But no. Five of the guards who had been sitting in back with us bailed, while the sixth guard shoved the other prisoner and me under the skinny wood bench that ran around the rim of the truck bed. We were still chained together, unable to run. Then the sixth guard jumped out and ran after the others.

The American stranger and I were alone in the truck.

I pulled my blindfold all the way up, inched out from under the pointless cover of the flimsy bench, and stared up into the night sky. Above me to the west hung the most beautiful, and most deadly, sight I have ever seen. A panorama of a thousand floating, glowing, orange balls of light hung suspended in the sky, the tracers of the 37-millimeter rounds firing all around us. The orange flashes revealed the silhouettes of some of the gunning placements in the ditches lining the road. They were *right there.* I could see their dark shapes through the open tailgate: the low mounds, the shoulders and heads of men, the outlines of faces, the long barrels of automatic weapons jittering as they sent ammo flying in a sort of *ratta-tat-tat-tat* but sharper, like a whip cracked by God. The bombs shook the bowels of the earth, closer, closer, closer: *boom, boom, boom.* Then an 85-millimeter *BA-BOOM!* The calling card of the devil.

One clue told me we were right at the center of the battle: those little orange balls melted ever so slowly, though in actuality they were moving fast and only appeared slow because they were traveling away from us. For each of the hundreds of orange balls I saw, there were several more rounds I did not see, chasing after them like a comet's tail.

I'd never seen anything like that from the ground before. I felt an intense thrill of adrenaline. Then I heard a plane pass right over us at about two hundred feet, the combination of a high squeal and a sound like an erupting volcano that vibrated every inch of me. The only sensation that might be scarier would be lying in the center of train tracks and letting a freight train thunder right overhead. I didn't even know that the military had aircraft that could fly that low at night. I thought I heard the pilot firing something like a 20-millimeter canon or rockets, but no bombs. The truck felt as if it might shake itself to bits. I thought it quite likely that this was the last moment before my death. If so, it seemed both a horrific and glorious way to go. If my goose was cooked, then I wanted to keep my eyes and ears wide open to the end. Perhaps for the first time, I felt how small I was, a speck of dust in the center of an infinite universe.

All it would take was one flare dropped near the truck and we'd be in deep shit. The pilots would see the truck in the middle of the road and the gunnery placements around us, and the next thing we would hear would be nothing at all—because we'd be dead before we knew what hit us.

Instead, the planes made their way back to the gulf, the firing died down, and my ears began to ring. I felt the chain that linked me to the other prisoner tug lightly at my ankle as we eased ourselves all the way out from under the bench and sat in the middle of the truck bed. We both stared at each other, face to face for the first time, no NVA in sight. Darkness had returned, but I could see the whites circling his irises so that his eyes seemed to gleam in the dark.

With a breathless laugh he uttered his first words to me, "Man, that was scary."

To me that was the most profound understatement of the century, and I barked a surprised laugh. "No shit." *Terrifying as hell* might be more like it, but I didn't want to waste words. This might be our only chance to talk, so I rattled off the questions I had been rehearsing in hopes of such an opportunity. "What did they do to you?"

"They beat the shit out of me." He shrugged.

I understood: there was neither time nor words to describe all they had done to us. "Yeah. Same here. Where'd they shoot you down?"

"Mu Gia Pass."

"What were you flying?"

"F-105."

The F-105 Thunderchief was an Air Force jet. The Thunderchief flew what were arguably the most dangerous missions over North Vietnam. F-105 pilots attacked targets in the Hanoi area at ten in the morning and four in the afternoon, every day for two years. I wondered how an Air Force pilot and a Navy pilot ended up in the same place.

"What're you doing here?" I asked.

"That's what I'd like to know," he said.

I had to chuckle. I had meant the question literally, but I suppose our brush with death had prompted him to ponder that question on a deeper level. There was no time to clarify, or to discuss what possible meaning or purpose had brought us here, because that's when the guards returned. We quickly turned away from each other, silent strangers again.

The soldiers solemnly took their places around us and the truck took off as if nothing had happened.

I replayed the bombardment in my head as we drove on. I estimated that if we'd had to make that trip to Hanoi four times, then one of those times we would not have made it.

I no longer stewed over whether they were taking me to Laos to execute me as a spy. Instead, I thought about the guy who had been chained to me during that night's life or death moment, and my thought was this: Damn, I forgot to get his name.

And I forgot to tell him mine.

7

HEARTBREAK HOTEL

After surviving the aerial bombardment, we turned north, but it didn't immediately hit me what that might mean. The sun rose, and one of the guards turned on a portable radio. He spun the dial through several scratchy stations before zeroing in on a clear broadcast of a catchy Vietnamese folk tune. A chorus of cheerful women filled the air, their voices accompanied by playful notes plucked on stringed instruments. I wondered why the Vietnamese military would allow its soldiers to carry radios and listen to anything they wished. What if they landed on an American station or a subversive South Vietnamese talk show host? Then the significance of the broadcast struck me: we must be approaching Hanoi. Where else would there be so many stations or such a clear signal? Our journey was almost over and I would soon find out what the enemy had in store for me at the end of the line.

Not so fast.

We stopped in another village for the day, bigger than any we'd seen so far; this was more of a town. Soon the hut where the guards placed us was surrounded by a mob, a crush of young people who screamed, shoved the bamboo structure, peered through the windows, reached under the walls, and spit through the gaps between bamboo poles. We sat helpless in the center of the floor while our guards stood around us and contemplated the crowd. I would not have been surprised if the people caved in the hut and beat us to death.

The guards exchanged significant glances before taking us outside to give the locals a look at us. This move could either calm the villagers or incite them to tear us apart. If it was the latter, I doubted these twitchy young soldiers would run interference. They were outnumbered and I couldn't imagine they had much incentive to fight for us. The crowd did not attack, but continued to shout as they milled around us, sometimes lurching forward in a threatening way like predators intimidating their prey. After a few minutes, the guards took us back inside. For the next few hours, it seemed they were waiting for something.

By siesta time, the crowd had dispersed. The village fell nearly silent while people napped or relaxed through the heat of the day. Our guards took advantage of the lull and snuck us down the path to another hut. The crowd didn't find us again.

At sunset the other American and I were loaded into a truck, chained, blindfolded, and carted onward. A dozen guards rode with us this time. Our ejection seats and pilot gear sat in the middle of the truck bed covered by a green canvas tarp. It was a mild night. Well after dark, I felt the road under us change from rutted mud, to rattling gravel, to smooth pavement. Not asphalt. Harder. Concrete maybe. The manmade smells of city life filtered into my nose. Around 1:00 a.m., dots of bright light seeped through the gap in my blindfold in rhythmic succession. Street lamps. Then a stretch of darkness that glowed dimly. Moonlight. Then more lamps, more moonlight, more lamps. I heard no other engines or tires. The roads seemed deserted.

I was startled to hear the cry of a train. Its shrill whistle fell to a mournful howl, creating the impression it was speeding to a far-off destination. I was startled because we had been told on the carrier that Vietnam's trains were unable to run since we had bombed all their railroad yards and significant stretches of track. How did we miss a target that big? I imagined a hundred cars clicking over the tracks, each filled with military cargo: weapons, ammunition, food, medicine, maybe trucks, maybe even troops.

The next thing I knew, the truck stopped and the guards unshackled us. When a guard removed my blindfold, I saw that the other American pilot was gone. We stepped out of the tailgate onto a lonely street in front of a broad low building. I had an impression of other buildings and streets in the distance, but the city felt empty. A few blocks away silhouettes of tall buildings reached toward the moon; we must have been close to downtown. Several guards flanked me and led me through towering double doors. Viking doors, I thought. The creaking sound of the doors opening and the clang of a lock told me this was the prison camp.

We entered a narrow concrete walkway with open sky above and the whitewashed concrete walls of two large single-story buildings on either side. The abundant institutional concrete told me this had always been a penitentiary. I wondered what the government did with all the Vietnamese criminals who would normally have stayed here, so they could make way for prisoners of war. I wondered where all the other POWs were, where the guards had taken my nameless comrade, where they were taking me. My bare feet slapped the cool concrete and the sound bounced off the walls on either side as we continued through the tunnel. We passed an opening to the right where I saw the silhouettes of plants and smelled the perfume of tropical flowers. Maybe this place wouldn't be so bad. I could not imagine Hell having a garden.

I heard a puny *tha-thunk, tha-thunk, tha-thunk.* The sound was familiar. We passed under an archway into a courtyard where two Vietnamese men played Ping-Pong in their t-shirts. It was an odd sight after 1:00 a.m., but I supposed the night shift must be boring.

They didn't even glance up when I passed, took no notice of me at all, just another prisoner. After two weeks of being watched by guards and stared at by villagers, it was almost a relief to step into a place where nobody seemed the least bit interested in me.

The guard leading our small entourage turned to me, said, "Wait here," and walked away. I looked behind me and saw that my other guards had gone. The guards playing Ping-Pong wore service revolvers strapped to them. Even if they hadn't, running was not an option. I was in a locked facility full of Vietnamese guards in the middle of Hanoi. There was nowhere to run.

I looked around. To my left stood a bicycle rack with a half-dozen bikes. Beyond that, shadows were interrupted by light bulbs on poles. After that, the dimensions of the prison grew hazy. I watched the Ping-Pong game until that *tha-thunk...tha-thunk...tha-thunk* hypnotized me like the ticking of a clock. Some ten minutes passed, and still the two players didn't look at me, even when one missed a shot and had to chase the ball.

A new guard appeared across the courtyard and waved me over to follow him, his gesture nonchalant, as if I were a houseguest he was showing to a guest room. We passed the bike racks and stepped onto another concrete walkway. To our left rose a building that at this hour looked like a funhouse maze of barred windows, doors, and passages. We walked about fifty feet, passed a couple of rooms, and stopped at the open doorway to another. "Go in there," the guard said.

I walked through a dim anteroom into a large backroom that had all the earmarks of an interrogation room. A small wooden table was pushed against the wall and a large desk trimmed with dark-blue felt dominated the corner. Atop the desk sat a motionless fan and half a glass of yellowish water, tea maybe. Behind the desk sat two straight-backed wooden chairs. Behind those, a metal bar sat propped against the wall. Did they use that to work over the prisoners? My eyes shifted away. In front of the desk sat a stool. The knobby plaster walls glowed sickly green in the light of a single bulb dangling from a high ceiling. Also hanging from the ceiling was a large black iron hook, the type I'd seen in butcher shops with sides of pork dangling from them. I stared at the hook and swallowed. I pictured myself hanging from it, a side of human meat.

"Stay here," the guard said, then left me alone again.

He left the door open. To me, that did not indicate lax security. Rather, it revealed that this place was so secure he didn't need to lock the door.

I looked up at the large window. Its wood shutters were thrown open to let in the night air. The window had no glass, only steel bars that divided the darkness into even strips. The view between them was of a concrete wall two feet away. I stared out the window until I heard the click of boots on concrete and saw the head of a guard bob past. Some twenty minutes later he came back the other way.

I kept turning to stare at the hook. Would they use it to torture me?

The stool did not tempt me. I had been sitting in the truck for hours. However, I desperately wanted to lie down. So I did, right there on the concrete floor. The hook hanging overhead was the last thing I saw before I dozed off, and it was the first thing I saw when approaching footsteps woke me. I jumped to my feet.

The man who entered appeared to be an officer. He wore a faded green military uniform and a pith helmet with a gold star on a red background, which made the open sandals on his feet appear all the more incongruous. He pulled a ditty bag from his shoulder and set it on the desk with the casual grace of someone who has done so hundreds of times. He opened the flap, pulled out a manila file, notebook, and pen, and tossed them onto the desk. He then sat at one of the chairs behind the desk, opened the file, leaned back, and gave me a droll smile. He was wall-eyed, so only one of his eyes met mine. The other seemed to look out the window, so I had the unnerving feeling he was simultaneously studying me and ignoring me.

"So, you have decided to visit our country?"

I did not respond.

He didn't stop smiling. I couldn't tell if that was meant to be friendly or intimidating. Between his off-kilter eye and the hook hanging over our heads, the smile sure didn't seem friendly. "What's your name?"

"Robert Wideman."

"Yes." He gave a firm nod, as if I had passed a test.

With that, he launched into a series of familiar questions: rank, serial number, birthdate, where I was from, what kind of plane I flew, what kind of payload I carried. After each correct answer, he either looked down at the file in front of him or gave a knowing nod and said, "Yes." So the file was a dossier on me. I wondered what it said.

He asked if the other interrogators had taught me how to bow.

"Yes."

"Show me."

I bowed for him the way I had been instructed, until my back was parallel to the floor.

He frowned and shook his head.

"They told me I'm supposed to bow ninety degrees."

"No, it is only necessary to bow thirty degrees."

I bowed again, just a slight incline of my back and head.

He nodded. "Okay."

It seemed he was trying to make certain that my previous interrogators had broken me and that he could now make me do or say anything he wished, a puppet on a string. I felt a twinge of irritation, but I was learning to keep my expression a blank slate.

The questioning lasted about fifteen minutes. Nothing new came up. Then he told me to stand in the corner with my nose to the wall and not move. The moment I turned my back he walked out. He left me there for about three hours. The only break in the monotony was the guard making his rounds past the window every twenty minutes or so. At one point, I felt faint, so I set the stool in my appointed corner, sat down, and fell asleep with my head propped between the two walls. Then footsteps woke me. I jumped up, put the stool back, and hurried back to the corner.

When I heard someone enter, I turned my head and said, "I'm awfully tired. May I lie down and sleep?"

It was a new guard. "No, you must stand in the corner."

He said nothing else, but left almost as quickly as he'd come. I disregarded his order, at least the part about standing. I sat in the corner, and again fell asleep with my head pressed into the place where the walls met. I slowly slid to the floor and passed into oblivion until one of Vietnam's infernal roosters woke me at dawn.

I lay staring at the first motes of sunrise until I heard stirrings, footsteps, and snatches of Vietnamese conversation. A guard brought me a meal at about 9:00 a.m. It was a lukewarm tasteless soup with potatoes and a few scraggly pieces of beef, but it tasted better than anything I'd eaten at Vinh; I found out that was the village where the NVA had held me the previous week. The guard also gave me a generous little loaf of bread and two small bananas. The bananas looked black—though maybe they were a natural purple—and I figured they were rotten so I didn't touch them. As it was, I could barely finish the meal. I had eaten so little for two weeks that my stomach seemed to have shrunk. Still, I ate as much as I could, determined to keep my strength up.

I was forcing down the last bite of bread, when air raid sirens screamed across Hanoi and a gong sounded across the compound. I scrambled to my feet and looked out the window, but all I could see was a large tree fanning over the deserted courtyard and a red brick building across the way. The place did not look nearly as spooky as it had the night before, except for the sudden stillness. It seemed everyone had taken cover.

I looked skeptically at the desk and didn't think much of my chances should I hide underneath it. The wood would serve as mere kindling. I gave up on the idea of cover, sat in the corner, tucked my knees against me, wrapped my arms around my knees, and waited. The sky filled with the roar of jet-engines, artillery, and bombs. The ground shook under me like a pounded drum. With each shaking breath, I told myself there was no reason for my terror. The American government must know where this prison camp was and would surely avoid it. Still, it was hard to convince my body it didn't need to run and hide, because this was the loudest raid yet.

About ten minutes later, the roar subsided, and I heard shouting around the compound. My ears had barely stopped ringing before another wave began. It seemed like the raids happened every thirty minutes. I learned later that the Americans bombed Hanoi a lot more than usual that day: May 19, 1967. Several hours passed before I saw another human being.

In the afternoon, a guard opened my door, blindfolded me, and walked me in a strange pattern: repeatedly turning, taking me up steps, down hallways, across courtyards, down steps, turning, turning, and turning again until I had no sense of direction. This was probably partly because the prison really was a maze, but I suspect he added extra turns and switchbacks to ensure I would have trouble concocting an escape route, as if I dreamed of trying such a thing. All I could see under the edge of my blindfold was the changing floor beneath my feet: concrete, steps, gravel, red tiles, more concrete. After about five minutes, I sensed walls closing in around me. I stumbled up a step I didn't see and into a room.

"Sit down." The guard pushed me into a sitting position on a raised platform. "Do not take off your blindfold." He walked out and shut the door with a heavy metal clank.

I sat waiting for the next instruction, not sure whether there was anybody else in the room, or whether the guard might return. A foul smell besieged my nostrils, a blend of human waste, ancient dust, antiseptic, and mildew. I peered under the blindfold at a patch of floor so dense with gray, gritty, moist unidentifiable filth that I lifted my feet, until it dawned on me I couldn't avoid the floor forever and I lowered them again. After several minutes of utter silence, I figured I was alone. Screw this, I thought, I'm not going to sit blind all afternoon in the waiting room to Hell. So I took the blindfold off. I immediately regretted it.

It was the grimmest moment of my life. What I was standing in was no room. It was a prison cell, about six feet by four feet, not much bigger than a coffin. Of course it was a prison cell; this was a prison. Even so, it was a shock. The walls had been whitewashed once upon a time, but that was clearly years ago. Now they were grimy and smeared with splotches of dried red liquid. Blood. And something else: iodine or Mercurochrome. The red and orange were also splashed on the platform underneath me. I jumped to my feet in disgust, and saw that I had been sitting on a narrow concrete bed that ran the length of one wall, opposite a matching concrete bed against the other wall. Each bed was about a foot and a half wide with a pair of locked iron manacles bolted to the ends. The beds were strewn with wads of bloody gauze and bloody medical wrapping tape. Neither bed had a mattress, pillow, sheet, blanket, or anything to offer even a pretense of comfort.

Atop one bed sat the broken bottom-half of an old broom with only a few bristles left behind like the last straggly hairs of a dead man. Underneath the other bed squatted a large white enamel chamber pot with blue trim. It had a lid on top and a handle for carrying. I suspected that was where the stench came from. Stupid me, I had to check: the pot was filled to the brim with festering feces and urine. I slammed the lid back into place and swallowed the urge to vomit.

Together the two concrete beds took up most of the floor space, separated by a skinny passageway about a foot wide. At one end of that passageway was a metal door, at the other a hole in the floor for urinating. Urine would run into a drainage channel that vanished under the wall. I imagined myself slowly going insane, babbling and pacing the six-foot-long passage from metal door to urine hole and back, with nothing to light my way but the sizzling filament of the single bulb dangling overhead. I could detect only the meanest hint of sunlight leaking in from a tiny half-boarded window near the ceiling. I stood on each bed, but was no closer to reaching the view out of that useless window. The ceiling was some twenty feet high.

I later learned the high ceiling was to make up for the lack of air-conditioning. Hot air rose and cold air stayed low. That explained the ceiling, but not why the *window* was so high too. I assumed it was intended to tantalize and punish the criminals inside.

The cell's metal door had a tiny barred opening with a cover that a guard could open to peer inside. Next to that, a paper with English writing on it was taped to the door. A title at the top of the page read: *Camp Regulations for American War Criminals Captured in Vietnam*. The paper explained I was a war criminal because I had committed "black acts" against the Democratic Republic of Vietnam (DRV). Interrogators had repeatedly informed me that as a pilot I was the blackest of war criminals because I had killed civilians. Of course, civilians were not my direct targets, but it was possible I had inadvertently killed a few with my payloads, and I believed that my plane had probably killed civilians when it crashed. Anyway, the paper said that as a criminal I was subject to the listed rules. I don't remember the exact wording, but the list included the following:

Criminals are under an obligation to give full and clear written or oral answers to all schemes and attempts to gain questions raised by camp authorities.

All attempts and tricks intended to evade answering further questions and acts directed to opposition by refusing to answer any questions will be considered manifestations of obstinacy and antagonism which deserves strict punishment.

The criminals must demonstrate a cautious and polite attitude to officers and guards in the camp.

When the Vietnamese officers and guards come to the rooms for inspection, the criminals must carefully and neatly put on their clothes, stand at attention, bow a greeting, and await further orders. They may sit down only when permission is granted.

The criminals must maintain silence in the detention rooms and not make any loud noises which can be heard outside.

All attempts at communication with the criminals living next door by intentionally talking loudly, tapping on walls or by other means will be strictly punished.

The criminals must go to bed and arise in accordance with the orders signaled by the gong.

During alerts the criminals must take shelter without delay. If no foxhole is available, they must go under their beds and lay close to the wall.

Criminals must not write on the prison walls.

Any obstinacy or opposition, violation of the preceding provisions, or any scheme or attempt to get out of the detention camp without permission are all punishable. On the other hand, any criminal who strictly obeys the camp regulations and shows his true submission and repentance by his practical acts will be allowed to enjoy the humane treatment he deserves.

Anyone so imbued with a sense of preventing violations and who reveals the identity of those who attempt to act in violation of the foregoing provisions will be properly rewarded. However, if any criminal is aware of any violation and deliberately tries to cover it up, he will be strictly punished when this is discovered.

In order to assure the proper execution of the regulations, all the criminals in any detention room must be held responsible for any and all violations of the regulations committed in their room.

I looked from the tattered list smudged with rust-colored stains to the cell closing in around me and said, "*Detention room*? This is a dungeon!" A laugh escaped me, the kind of laugh that might slip out of a man if he died and Satan walked him to a tiny cell and said, "Bad news, buddy..."

I turned in a circle, trying to grasp reality. Like it or not, I would be stuck indefinitely in this hellhole where the geckos clinging to the walls changed color to blend in with dirty whitewash and dried blood. Up until my plane had crashed I had not seen any blood in this war, except during my off-duty time back home.

One night in Lemoore, when Pat and I had been married just a month, she was in the kitchen slicing an avocado and I was in the bedroom when I heard her call out in a high, quivery voice, "Bob? Bob?" I ran into the kitchen and was shocked to see blood squirting two feet from her hand to the shelves above the sink. A thick, bright arterial spray. She had cut herself deeply in the fleshy web between her index and middle fingers. I wrapped her hand in a towel, covering myself in blood in the process. It took too long to get help from the operator so I drove Pat to the hospital, running every red light on the way. I wanted to be supportive, so I sat with her in the emergency room while the doctor stitched her up. I watched the needle go through her bloody flesh and felt my own blood rush to my feet until I almost passed out. Just look at me now.

A lot of pilots back on the carrier pinned up photos in their lockers and above their bunks, of wives, girlfriends, and the movie stars of the day. The leggy Ann-Margret was a popular choice. But if any of those ladies had appeared before me at that moment—hell, if my wife had appeared before me—I would have only asked one question: "Can you help me get out of here?"

At about four o'clock the sirens and gong sounded again. I had read the rules: *During alerts the criminals must take shelter without delay. If no foxhole is available, they must go under their beds...*I dropped to my knees, ready to scramble under the bed, but when I peered underneath I saw a tangle of cobwebs. I weighed the likelihood of American bombers striking a known American POW camp against the likelihood of being bitten by a poisonous spider or diseased rat. Other clumps of unidentified detritus nested among the spider webs. There was barely enough room to squeeze in anyway. What's more, if a bomb hit, that concrete bed would just be one more piece of debris to collapse on top of me. I got up and sat down on the bed to wait it out.

I held my hands over my ears, bracing for the punishing roar of yet another air assault. I was startled by the sound of the metal flap on my door flipping open. I turned to see a tall, skinny Vietnamese man glaring through the barred square in the door. He pointed under me and grunted loudly, indicating I must climb under my bed.

I didn't move. "It's dirty."

He pointed under the bed again. I didn't get why he was so insistent. It was *my* life. What did he care if an enemy lived or died? The NVA might consider me an intelligence asset or useful as leverage, but surely this one guard, just another cog in the wheel of war, didn't care about that.

I made a calming motion with my hand, trying to appear reasonable. "I'll take my chances. I'm not going to sue you if I get killed. Don't worry about me."

I didn't know if he understood me, though under the circumstances my tone and gestures seemed clear. In any case, he was brooking no disobedience. He opened the door, stood over me, and jerked his arm in an irrefutable gesture: Get under!

I pictured unseen creatures crunching under me, stinging, biting, bursting. I shook my head hard. There had never been anything I was so sure I did not want to do.

Before I knew what was happening, he slapped me full and hard in the face.

I dropped and crawled beneath the bed, my skin shrinking from the floor where I felt cockroaches or other insects scuttling away. It was even filthier than I had imagined, coating me in the bodies, dust, and waste of God knew how many vermin, alive *and* dead. The guard stalked out. I didn't dare climb back out, lest he come back and do worse than slap me. The ground rattled my ribcage while the apocalyptic sounds of war filled my ears, and the crypt-like smell of the offal around me overwhelmed my nostrils.

I stared up at a series of squares some previous tenant had scratched into the bottom of the concrete bed just a couple of inches above my face: a calendar, with the name D. Luna scratched next to it. The calendar was six months long and D. Luna had scratched out every single day in the calendar. I wondered if he had been in this room for six months or if he had simply given up on keeping track, if the guards had moved him to another room in the prison or if he had died.

When the all-clear sounded, I shot out from under that bed as if I'd been shot from a cannon.

I paced the center walkway and thought about D. Luna's calendar. I imagined Alexandre Dumas *Count of Monte Cristo* pacing in his dungeon on the island prison of Château d'If, scratching his own calendar into the wall every day for fourteen years. Edmond Dantes had been framed for a crime he did not commit. I identified with Dantes. For years, nobody knew what became of him. Did my family know where I was? Dantes ultimately escaped the Château d'If, christened himself the Count of Monte Cristo, and achieved revenge.

But that was a novel, I told myself. This is 1967 and that kind of stuff just does not happen anymore.

Nonetheless, I immediately dropped to my knees, clutched my palms together, and prayed to the God I had abandoned nine years earlier. "Dear Lord, I know I've lost faith in you, and I feel like a hypocrite praying now. But if there's any way you can help me out here, I'd really appreciate it." I still doubted anyone was listening. On my knees, hope felt as hard to reach as the tiny rectangle of light twenty feet above me.

From all I had seen since I'd crashed, the DRV was much stronger than our superiors had told us. North Vietnam did not appear remotely ready to surrender. Would America? We never had before. I might spend years in this prison. What sort of man would I become if I remained trapped in this small cell for the rest of my days?

A strangled scream shot across the prison. I jumped as adrenaline surged through me. I looked out the flap in the door, which the guard had left open, and saw a prisoner walking into a shower stall. He was a mere twenty-five feet away, but he may as well have been on the moon. If I called out, I would surely be punished. Hungry for contact with another American like me, I satisfied myself with staring at his head as it bobbed past. But his head was unlike mine: short grey hair circled a bald spot, calling to mind a medieval monk. To me, he looked a hundred years old. He was probably 40. Too old to be a pilot, I thought. I was wrong about that. I was among the youngest pilots at this prison I would come to know as The Heartbreak Hotel.

In desperation, I reached for the one object in the room that remotely suggested civilized life: that pathetic corpse of a broom. Although it had only a few bristles left, I set each one to punishing work, scraping out every cobweb, spider, insect carcass, rat dropping, and human skin cell I could. With that one small act of control over my surroundings, I began to hope I might maintain a modicum of control over my sanity—however long I had to stay here.

Lucky timing. I had barely finished sweeping when the air raid gong sounded and I dove back under the bed. I squeezed my eyes shut so I would not have to stare at D. Luna's calendar again. But it was too late. I knew it was there.

8

SHAVE AND A HAIRCUT

The air raids decreased to twice a day and became another routine of prison life. They arrived at almost the same time as my meals: 10:00 a.m. and 4:00 p.m. The meals offered almost no variety: weak tea or tepid water, thin lukewarm soup, bitter boiled greens, piles of tasteless white rice, occasional potatoes, and a few slivers of beef or chicken or, more frequently, pork fat with the pig's skin and hair still clinging to it. In the prison, I enjoyed one regular addition to my daily menu that seemed out of place with the rest, until I remembered that Vietnam had spent six decades as a French colony: small loaves of French bread—not rolls but whole loaves!—crispy outside, soft inside, sometimes not even cold or stale but actually hot and fresh. Although I could have used a few extra pounds, I couldn't finish all the bread they gave me.

One day a guard came into my cell with a meager allotment of personal supplies that seemed like untold bounty after weeks of deprivation: two short-sleeved shirts, two long-sleeved shirts, two pairs of long pants, two pairs of shorts, a bamboo sleeping mat, a mosquito net, a ceramic teapot with a handle and lid, a white-and-blue enameled metal drinking cup with a lid, an aluminum spoon, a bar of soap, and to my great surprise, a toothbrush and toothpaste. My survival school instructors had given me to believe that my teeth would rot in prison. I was relieved to find that this would not be the case. Still, it was difficult to be grateful to my captors for these meager comforts. My aluminum spoon was made from the wreckage of downed American airplanes. It struck me as an act of disrespect, an insult, to eat with it each day.

While my teeth were soon clean, I couldn't say the same for the rest of my grooming routine. My hair was a tangled mess and my face was covered in an unruly beard that itched like mad. I longed for a shave and a haircut.

Of all the items in my new kit, the soap was my favorite. I set my sights on little comforts like that, which distracted me from my homesickness and fear about the future. Ordinary hygiene felt like a luxury. My first days as a captive had been disgusting, but that changed at The Heartbreak Hotel, where the guards took me to the shower every day except Sundays.

The shower was in a cell across from mine, which meant precious moments outside my claustrophobic confines. Walking from my cell to the shower, I saw that mine was among ten prison cells, five on either side of the outdoor hallway. I had no idea how many prisoners were in those other cells, because I never heard them. Maybe a rustle or murmur now and then, nothing more. If I had neighbors, at this point they were following the rules. No Americans had tried to talk to me since I'd arrived. In fact, the only sorts of communication I heard that first week were orders barked by the guards.

The days were so long and lonely, I soon looked forward to the duck-like calls of the geckos, talking to each other as they clung to the walls amid roaches as big as they were. I stared at those curvy little lizards long enough to witness them changing color like chameleons as they moved from the bloodstains to the whitewash to the green-painted metal doors. Sometimes one might mount another to mate. It was my only source of entertainment.

I almost welcomed the day a guard came into my cell and signaled for me to get dressed, pointing to my long pants and long-sleeved shirt. I was clearly going somewhere where it was important that I look presentable. I didn't know if this was good news or bad, but at least it was a change. He led me back to the room where I'd spent my first hours in this prison, the one with the green knobby walls and the hook. He gestured that I should enter, then stationed himself outside the door.

The wall-eyed interrogator sat behind the desk. Crooked Eye—as the prisoners came to call him—gave me a sly look and asked what I thought of the Bertrand Russell War Crimes Tribunal on Vietnam. I had never heard of it. He said it would be getting underway soon and described it as similar to the Nuremberg Trials. He mentioned Jean Paul Sartre, who was a supporter of the tribunal's purpose: to uncover whether the United States was engaging in crimes against humanity in Vietnam. I had no idea who Russell and Sartre were. Crooked Eye thrust a Vietnamese newspaper under my nose and pointed to a photo of Sartre standing at a podium surrounded by men. They all had long hair, beards, and bushy sideburns. To me they looked odd, different from the clean-cut men I was used to, though all I said to Crooked Eye was, "So they're communists."

He told me they were trying President Johnson as a war criminal. "Wouldn't that be something if they found him guilty?"

"Yeah, I guess so," I said.

He shook his head, whacked the newspaper against the desk, and tossed it aside in disgust, as if I were too big an idiot for intelligent political discussion. With that, he launched into a more detailed interrogation than he had at our first meeting. It started almost like a job interview, not relaxed but with assurances that so long as I answered his questions we would get along fine. He sat behind the desk and opened a notebook with a black-and-white speckled cover. Just as my other interrogators had, he started by asking about my family, where I grew up, whether I had children. Just as my other interrogators had, he then started the military questioning with "What was your last target?" I had given the answer to that one after my first interrogators broke me, so I saw no reason to hide the truth now.

Then he asked a question I had not yet heard. "How many missions do you have?"

"Ninety." I had flown 134 missions, but I figured fewer missions would make me seem less of an evil war criminal to this guy whose country I had bombed. I didn't want him to think I was a big war hero but just another cog in the machine. I kept my face still, testing him as much as he was testing me. His answer would tell me what life would be like in this prison where I expected to spend the duration of the war

"No! Not enough!" His response was shrill, and his crooked eye rolled in the socket. "How many missions do you have?"

"I've flown ninety missions, sir."

"You lie."

He repeated the same question and I repeated the same answer over and over again, his impatience growing each time I repeated the number "ninety." He seemed to be struggling to maintain self-control. He glared at me, chest heaving, and pointed to the corner of the room.

"Now you must stand in the corner until you are ready to tell the truth." I walked to the corner and faced it. He walked out.

I don't know how much time passed—a half hour, an hour—before a guard came by with my meal and set it on the floor behind me. That meant it was about four o'clock. When he left, I turned to glance at my tray of the usual: thin soup, squash, a French loaf, tea. I didn't dare move toward the food, in case tantalizing me was part of my punishment and someone was watching. I turned my nose back to the corner.

Then I heard another set of footsteps accompanied by Crooked Eye's now familiar voice, "Okay, you can eat, but you must eat standing at the wall."

I chewed slowly, in no hurry to be alone with that corner again. It was no use. I had time to get to know every knobby lump of that green paint. I found faces in the shapes, saw my friend Danny and the model airplane we had doused with chemicals from the chemistry set I got for my fourteenth birthday. I set it on fire and threw it out the bedroom window. The plane landed on the grass in flames just as Dad's peach 1957 Plymouth pulled into the driveway. "Oh man, am I in trouble," I groaned. My father gave me what-for: "You could have set our house on fire!"

I thought of my 134 missions, my plane crashing into several bamboo huts, bursting into flames. Oh man, am I in trouble.

Another thirty minutes passed before Crooked Eye returned. "Turn around."

I turned.

"How many missions?"

"Really, it was just ninety missions," I told Crooked Eye.

"It's not true." His voice was calm, too calm. "If you do not confess correctly, I will put you in a small room with your hands tied behind your back."

In survival school, the instructor had tried to prepare us pilots for such a moment by making us spend an hour crammed inside a black box about three feet long by one foot wide. Now I knew that preparation had been meaningless, because it did not start with uncertainty about what was going to happen. Uncertainty was part of the torment. How small would *this* box be? How long would I be in it? What would they do to me when I got out?

Crooked Eye gave me plenty of time to think about it, leaving me to stand in the corner again.

Twenty minutes later he returned. "Did you have ninety missions?"

"Yes."

"Okay, now you're going to go back to your room." What happened to the box? I was ready for the box. The switch was disconcerting. Was I really going back to my cell or somewhere else? He called a guard into the interrogation room, muttered a few terse words in Vietnamese, then turned his back on me and faced his own wall.

The guard marched me back to my cell, where he instructed me to take off my clothes and lie on the cement bed. Then he locked my ankles in the stocks. Instead of inserting my ankles straight into the metal restraints, he first crossed my legs and then locked them up. At first, this was merely awkward and not exceptionally painful in and of itself. But when it came time to sleep, I felt a natural urge to roll onto my side to change to a more comfortable position. That only made things worse because when I rolled over my big toe was compressed painfully against the concrete bed. I returned to my back, still sore from my recent injuries. I wouldn't have thought it possible to feel any more trapped in that room until I spent a long night in those stocks. I barely slept.

It was difficult to tell when night ended and day began, because the room's single pale white bulb was left on at all hours, day and night, and the only reliable sunlight was trapped behind the metal flap that covered the cell door's window. My whole consciousness focused on the place where the two sharp bones of my ankles pressed into the metal stocks, until a guard opened the door flap and spoke a single word.

"Talk?"

"Yes."

The guard entered, unlocked the stocks, and instructed me to again dress in my prison-best. I rose to unsteady feet and complied, slowly, every muscle almost as stiff and sore as that first day I had plummeted to earth. He returned me to the knobby green interrogation room, the dangling hook, and Crooked Eye. I stood at attention, hands behind my back

Crooked Eye gave me a sardonic smile. "Good morning. Did you sleep well?"

"No." I saw no percentage in acting tough.

He again opened his notebook. "How many missions did you have?"

"You're not going to believe this, but I only had ninety missions." My own answer surprised me. Did part of me still believe I could win this game?

He surprised me too by chuckling as though we were sharing a joke. "No, it is not true." But instead of asking the question again, he moved on. "What was your last target?"

I stared straight into his one good eye. "You should know my last target. Your guys shot me down over it!"

He did not react. Instead, he asked, "Who is the deputy commander of your air wing?"

"I don't know. I was new and I wasn't there long enough to learn the commander's name."

I wasn't getting away with it. "You have a bad attitude. Get on your knees."

I lowered myself to my knees.

"You must stay on knees until your attitude improves." He continued to ask me questions while I remained on my knees, questions I had already answered many times before.

At first it didn't seem like much of a punishment: kneeling. But the discomfort of bone against cement grew with each minute, until his questions were nothing but a buzz in my ear and I was reduced to a pair of throbbing knee bones. I tried to shift my legs just a hairsbreadth, hoping he wouldn't notice and strike me. He again asked who the deputy air wing commander was. I gave a false name.

With a look that told me I wasn't fooling him, he slid a paper and pencil across the desk. "Write down the names of the pilots in your squadron."

"It might take me a while to remember them all." I was stalling, wondering if there was some way that giving those names might get those pilots shot down.

He gave me a measuring look. "Take your time."

He walked out, and a guard walked in to watch me. I pulled the stool up to the desk, sat down, and made up a dozen names. My hand shook as I wrote. I handed the list to the guard to take to Crooked Eye. He returned to the room, set the paper in front of me, and splayed his fingers across the names like the legs of a spider.

"No. You must do it again."

This time he stayed and watched the progress of my pencil as I wrote a list of actual people who had once been in my squadron but not anymore. I pushed the paper across to him.

He glanced at the first few names. "No, not true."

I took a long breath and exhaled slowly, weighing the consequences of telling the truth. I realized he already knew the truth or he would not be so confident in dismissing my phony lists. I dragged the paper back toward me and wrote an actual list of the pilots in my squadron.

He looked at it, nodded, and pushed it aside, as if it were less important than a grocery list. Then he looked up at me. "How many destroyers do you have in the gulf?"

"One hundred." We only had three.

"How many route packages does your Navy have?"

I lied.

"How many route packages does your Air Force have?"

I lied again.

"What is your friend-or-foe transponder frequency?"

I gave him a bogus frequency.

I figured he would not know the answer to that last one because the frequency had been changed recently. He let me lie to his questions without interruption. I began to rejoice inside, mistakenly believing I had beaten him.

Then, when I finished, he slammed his notebook shut and threw it so hard it slid across the desk and almost fell into my lap. "You lie! Now you will be punished!"

My stomach felt as if I had just swallowed a bucket of ice. My bladder constricted. I tried not to look up at the hook.

He called in two guards and gave them instructions in Vietnamese. One of them left and returned carrying a heavy metal object. The other pinioned my arms behind my back and clasped my wrists together while the first one locked the metal object around them: manacles.

Manacles are not like handcuffs. Handcuffs have a chain between the cuffs that allows leeway for movement of the wrists and hands. These manacles had a figure-eight shape designed to force the wrists against each other. The guards used a screwdriver to screw the device into place, immobilizing my hands. Then they tied my elbows together with a parachute shroud, pulling on the ends until the elbows touched. The pain in my shoulder joints was excruciating, almost as bad as the pain I'd felt when the guards in Vinh had cinched my elbows with rope. The difference was meaningless. Once pain reaches an unbearable level, it all seems the same. I could not catch my breath. If I felt trapped before, I now felt the utter panic of claustrophobia. My entire body might as well have been squeezed into the box Crooked Eye had promised the day before.

The guards forced the end of a small hand-broom between my teeth. I bit down so hard, it seemed I might split it in two. The taste of wood filled my mouth as I flopped face-forward onto the floor. Crooked Eye leaned close to my ear and whispered, "Don't cry. If you cry, I will leave you like that and your arms will fall off." Then he followed the guards out.

I didn't cry. I feared maybe he meant it: that if I cried he would leave me like this until I developed gangrene. So I just lay on the floor writhing uncontrollably, my body seeking relief but unable to find it, my brain hoping to pass out but unable to will it. Talk about a tough negotiating position, I thought. How long will they leave me like this? An hour? Two? Three? Pure pain reminded me of what I already knew: it wouldn't take long before I would tell Crooked Eye anything he wanted to know.

Surely he knew that too, because it seemed like only twenty minutes—though not a minute too soon from my perspective—before he stood over me again, ready for me to revise my answers.

A guard removed the manacles and the pain vanished so fast it was hard to believe it had ever been there. Guilt replaced pain. Had I given up too easily? The guard pushed me back onto the stool while my interrogator eased back into his chair.

He leaned forward. "Why did you lie?" His tone was that of a disappointed father.

So I adopted the tone of a penitent son, much as I used to when my actual father was angry. "Because I'm stupid!"

What Crooked Eye did next made me feel as if I really were stupid. He opened his notebook and read aloud the same half-dozen questions I had failed to answer correctly, but instead of waiting for me to answer them, he read the answers aloud, one after the other. He knew my last target, the number of destroyers we had in the gulf, all the route packages we flew, and our latest friend-or-foe frequency—down to the decimal point. I tried not to show my feelings, but his smug expression told me he knew I was stunned. There was nothing that our soldiers, pilots, and sailors on the front line knew that the North Vietnamese didn't know. All my efforts to stymie the enemy, all the agony I had endured to protect my countrymen, had been for nothing. The prisoners before me had given up everything.

He resumed questioning me. I didn't bother lying again. I had no energy left to lie. This is my life now, I thought.

I felt a little nervous when he asked me the technical name for my plane's attitude indicator, because I honestly couldn't remember. AJB-something. I searched his eyes. Did he know? Of course he knew. These men knew everything. The unsettling thing was, if he thought I should know too then I was screwed. I couldn't face those manacles again. So I guessed.

"AJB-5." I was sweating bullets.

He looked at his notebook. "Would AJB-3 be it?"

I nodded. "Yes."

After I answered all his questions, he sent me to my room without further punishment. For the remainder of the day, I sat on my cement bed, stretching and twisting in a futile effort to ease the pain in my shoulders, twiddling my thumbs in a futile effort to still my thoughts in the crushing silence. I spent the next few days in my cell, until boredom became its own form of torture, a manacling of my mind. Two meals and two air raids a day were the only break in my solo routine. Then one day a guard came into my cell with a razor and a bowl of water.

Be careful what you wish for.

The guard had me lather my face with my cheap bar of soap and the water, which was cold. Then, without bothering to clip my dense beard, he shaved me, each movement rough and careless.

The razor had two blades, but both were dull and rusty. I must have been tenth in line for that excruciating razor. He sliced my face up more in that single shave than I had in all my shaving years. At least the haircut didn't hurt, much. After he left, I ran a palm across my raw face, and it came away bloody. I worried I might contract hepatitis or some other heinous blood-borne disease.

If I did get sick maybe there would be an upside. Maybe they would send me to the infirmary, where I just might get to talk to another person. One who would not make me pay for the things I said.

9

ROOMMATES

I found out later that my cell was Room #8. We prisoners just called it The Rotten Cell. I wasn't the only one who ever stayed there. Other pilots had spent stir-crazy time in The Rotten Cell before me, and more would end up there after me. Did they listen to the train too as it rolled through the city every night? My heart lurched with the train's howling whistle as it stretched its way out of Hanoi while I remained trapped. There was too much time to think, about my wife, my mom, my brother, my dad, and when I'd ever see any of them again.

Four years. That's what I told myself. Three seemed too optimistic, five too desperate. Pacing my six-foot path between the beds, I had a lot of time to consider the question. I didn't figure Americans as the type to accept defeat, and it was becoming clear these North Vietnamese weren't the type either. This could take a while. Still, humans had only so much stomach, or money, for war. World War One lasted four years. World War Two lasted six, but the U.S. was only in it for four. The Korean War lasted three. Four seemed about right.

About two weeks passed. Then one day a couple of guards walked in. One of them pointed at my clothes and personal items. "Roll up. Go to another camp." There was no time for dismay. I placed my toothpaste, toothbrush, plate, cup, and utensils atop my clothes and rolled them up in my bamboo bed-mat. Then the guards walked me through the compound to the main gate. Instead of opening the gate, they blindfolded me and we turned left. I never heard a gate open. I felt certain we never left the prison.

Instead they hurried me through a narrow maze with concrete walls on either side, scurrying through turn after turn until I was completely disoriented. Then they took my blindfold off and I was standing next to a long low building similar to the one I had been staying in, outside the door to a room similar to the one I had been interrogated in. One of the guards gestured for me to put down my gear and wait. He walked into the room, and came out two minutes later. "Go in," he said, pointing.

Inside the room, a uniformed Vietnamese I had never seen before was sitting behind a large wood desk. He had three stars on his shoulders.

He gave me a cheerful smile. "How's your health?"

I was puzzled. No Vietnamese had asked me that particular question before. "Fine."

"Are you getting enough to eat?"

"Yes."

I would come to know this guy as *The Fly*. But on this day, he was simply the first person I'd seen with a remotely friendly face since Mr. Easygoing in Vinh. "Do you want a roommate?"

It was hard not to smile back, though I refrained on principle. "Of course."

He reiterated the camp regulations: "You must not talk to other prisoners. You must not tap on walls. You must keep silence in your room..." This went on for a while, but I wasn't really listening. I was still trying to keep from looking too excited at the prospect of a roommate, worried that this officer might decide it was clearly too much of a treat for a POW and take it away.

He told me to sit down. I sat in a straight-back chair set against the wall. He left the room and reappeared about ten minutes later.

"Okay," he said and gestured for me to follow the guards.

The guards blindfolded me again to lead me to another building and more passages, before they pushed me into a room and once again took off my blindfold. There, standing in front of me, was the first white man I had seen in two weeks: the American pilot who had ridden with me in the truck from Vinh.

He too had endured a shave and a haircut since I'd last seen him, and it looked as if someone had used the same razor on him that they'd used on me, *after* a few more guys used it in between. His face was gashed deeply in a few places and scattered with globs of dried blood and blotches of whiskers. What a mess. His hair had been short when I met him, just a little messy from his ordeal, but now he had a flattop that let the unhealthy, pale, red-mottled skin of his scalp show through. I suppose I focused on the shave and haircut because I didn't want to take in how much weight he had lost. How could he look so much thinner in just two weeks? His eyes were sunk inside dark hollows. Did I look like I'd been through the same wringer? I hadn't seen a mirror since the air carrier, had only caught my ghostly reflection in passing windows. I never paused at those windows for a closer look, partly because the guards might see it as another excuse to punish me, partly because I feared seeing what I now saw before me.

Our eyes said it all: we had never been happier to see another human being as we were to see each other at that moment, while at the same time we were dismayed to see what we were seeing. However, we didn't smile or frown or say a word. I put my things down and turned to face the guards, who motioned for us to give our expected bows. I hesitated.

I had been bowing to interrogators and guards for weeks, every time I entered a room, every time I left a room, every time they entered a room, every time they left a room. But I had never bowed to a Vietnamese guard in front of a fellow American. He was going to see me lower myself to them, and I really didn't want him to see that. Then I realized he was expected to do the same. We exchanged a self-conscious glance, then averted our eyes. In unison we bowed. They closed the door and left.

We waited until the footsteps faded, then turned to grin at each other and pump each other's hands with all the enthusiasm of people who have not been touched in kindness by another human soul for a month.

"Good to see you, man!" I whispered. "I didn't think I'd see you again. Robert Wideman. What's your name?" "McCuistion," he whispered back, "Mike McCuistion. Glad to meet you."

"Hey McCuistion, this is going to sound crazy, but can I ask you a question?"

"Shoot."

"Who was *The Shadow*?"

He considered this for a moment, then gave me a baffled look. "I don't remember."

"It's been driving me nuts."

We shared a chuckle, as if he knew just what I meant.

The cell was a shade larger than the one I'd been in before, but not by much. Maybe eight feet square with two sets of bunk beds on either side. A narrow T-shaped passage led between the beds and alongside the far end of the room. I was relieved to note that the beds were wood instead of concrete and that the walls were green and only green, with no decorative splatters of blood or iodine. There was still the stench of chamber pots, but I was almost used to that. McCuistion's bed was already laid out, so I got to work on mine. I unrolled my bamboo mat atop my bunk and arranged my personal items at the foot. No more stocks. I rolled up all the clothes I wasn't wearing and put them at the head of the bed. I still had no pillow.

When I was done, I sat on my bunk and Mike sat on his and we appraised each other again. I was still shaken by what I was seeing in the closest thing I had to a mirror.

"How old are you?" he asked.

"Twenty-three."

"Oh shit."

"How old are you?" I echoed.

"Twenty-nine."

I didn't tell him how he looked, but it must have shown on my face.

"That bad, huh?" he said. "Well, don't gloat, you look like a wreck too."

I stifled a laugh, the first genuine amusement that had bubbled up in me since I'd taken off from the ship only a month before. It wouldn't be the last time Mike made me laugh. If he had been a college roommate, we would have been instant best friends. This went so much deeper. Our need for camaraderie was intense.

We rushed through our whispered military histories, each of us eager to solidify a bond with the only other American we had seen since our capture, maybe the only one we would ever see again. He confirmed the high risks faced by the Air Force's F-105 pilots at Mu Gia pass, one of the narrowest, most heavily armed passages along the Ho Chi Minh Trail.

"We lost a lot of planes out there. A real shame: best airplane I ever flew and I only got seven missions." He smiled and shook his head.

I studied the lines in my palms. I could barely hear my own voice when I spoke again. "So...did they break you?"

"Hell yeah, they broke me. Of course they broke me. Just like they broke you—you don't have to tell me, I know. You been feeling bad about that, buddy? You gotta let that shit go. Geez, they can really hurt you, you know, if you hold out too long."

I felt so much relief it was as if some part of the manacles and shackles had still been locked around my arms and legs until that moment, and he had just turned the key and released me. "I can't tell you how glad I am to hear that. I mean, I held out as long as I could..." I choked on the words and stopped.

He nodded with a knowing look that made him seem even older. "I know, man. Don't sweat it."

"What'd they do to you?"

He described all the same things that had happened to me, with a terrible addition. He didn't break when they manacled him, so they also locked shackles around his ankles and slid a ten-foot bar between them. They then tied one end of a parachute shroud to the manacles on his wrists, passed the rest of the shroud over one shoulder, down to the bar between his ankles, back over the other shoulder, and down again to the manacles, until he was doubled over. Then they pulled the shroud so tight that his nose touched the bar between his ankles. At that point he still refused to talk, although he could not stop the tears from rolling down his face.

"That's when Crooked Eye said, 'Just think Cuistion'—he could never get my name right—'Just think, Cuistion,' he says, 'this is but a speck in the annals of time.' I tell you, that was one big speck. I still wouldn't talk, so the guards flipped one end of the bar over until I landed upside down on the concrete floor, came down hard, and everything sort of compressed together. That did it. 'Untie me and I'll tell you anything you wanna know.'" Trouble was, before they could untie him, they had to flip him back over, which bounced everything together again. So he got a double-whammy for his trouble.

"Jesus."

He shrugged. "The stuff you went through was just as terrible. Unbearable pain is unbearable pain, right? Anyway, I resigned myself to the fact that I was going to wind up telling them what they wanted to know, regardless of how much I wanted to hold out. Once I realized I couldn't take it, there didn't seem any point to holding out. I was such a low man on the totem pole, I figured any intelligence I had probably wouldn't be important anyway. I mean, I'm not all that intelligent." He smirked, and when he did I could see he really was only 29.

"I can't tell you how grateful I am that you told me that. I can't tell you." At that moment I knew I was this guy's friend for life, no matter what might happen after.

"I'm glad my torture can make you feel better." This time we both stifled chuckles. Not that we thought it was funny—that would never happen—but we mutually understood something that could not be explained anywhere but in this place.

We talked for hours, making up for all the time we'd spent alone. He told me that back home he liked to golf and drag race. I told him about my new motorcycle. I had given Pat a ride right after I bought it, and when I stopped at an intersection and then gunned the engine, the bike sped out from under us and dumped us in the street. Mike and I had a good laugh over that one. I also told him about my dreams of becoming an airline pilot. He said he was career military but sometimes thought about commercial flying when he retired—if he got that far. I told him Pat and I had gotten married just nine months before, but hadn't seen each other in four months. He told me he'd been married and divorced, was now married again and had a couple of young kids.

"Two kids?" I blurted. "Wow, I'm sorry. I mean, that must be tough."

"What about you and your new wife? That must be tough too."

I wagged my head, acknowledging without saying that there really was no positive angle on being a prisoner.

We were still talking as I drifted off to sleep, not wanting to let go of the conversation, not knowing when my companion or I might be carted away. For the moment, I was no longer alone, and I drifted off feeling almost happy. Almost.

Talking with Mike made me think of my brother. I remembered Richard and me doing so much side-by-side, sometimes talking for hours about nothing or everything, sometimes not needing to talk at all. Together we walked to muddy Casanovia Creek with our fishing rods to catch a few measly fish, or wandered the woods with my wooden bow and the arrows we made together in the basement. One day we doused the arrowheads in kerosene and lit them with matches because we thought that shooting flaming arrows would look cool. One arrow set a bush on fire. We tore off our jackets to beat at the upper branches and stomped out the lower branches with our feet. We never told anybody. If Dad found out, who knew what he might do? Richard and I didn't need to discuss that decision. We shared a room, a house, a brotherhood. We just knew.

McCuistion could not replace my brother, but the possibility of finding a brother-in-arms in this dismal place lifted my spirits. Mostly, it was comforting to have someone to talk to who understood some part of what I was going through. Someone who just knew.

The next morning the guards let Mike and me out to bathe. They walked us to a bamboo stall that held a couple of brick cisterns with faucets, a place where we could take off our clothes and wash them, then soap ourselves and rinse off with buckets. There were about a dozen stalls lined up, and we could see the feet of other prisoners, but nobody dared speak. Mike and I didn't even speak to each other. It was forbidden. Still, it was better than my solo stint in The Rotten Cell.

The night after that, the door opened again, and there stood the hundred-year-old prisoner I had seen heading to the showers when I was still stuck in The Rotten Cell. Up close, I could see he wasn't that old, but he looked even more like hell than Mike. The bald spot in the center of his scalp was scabbed and bruised, one eye was completely red from a burst blood vessel, and his face was a half-shaved mess with angry patches of beard. He walked hunched over and shaking. He could barely carry his bedroll, which looked as if it were going to pull him over. I wanted to take it from him, but I didn't. I didn't want them to use my help as an excuse to punish him.

The three of us avoided each other's eyes and turned to bow to the solo guard; I guessed our new roommate must be too weak to need two guards. Then the door closed, and Major Dick Vogel introduced himself. He was only 35.

I don't know if older prisoners evoked more rage in our captors, but it seemed that Dick had suffered the worst punishment of the three of us. In his case, when they tied his arms together they also wrapped a rope around his neck and tied it to the rope around his arms, so that when he relaxed he would fall forward and the rope would strangle him. To avoid strangulation, he had to remain sitting up straight, which sent more intense pain shooting through his torso and arms, making him more likely to pass out. He felt terrified he might choke to death when the guards left the room. Dick admitted what Mike and I had already admitted: he had succumbed under the pressure and "confessed correctly."

I knew we didn't share the stories of our brushes with torture to impress each other, horrify each other, or receive pity. We just wanted to share this thing that nobody else could understand, except the other American pilots behind the other anonymous doors up and down the hall. I would learn later that Mike and Dick were exceptional men.

Still, neither of them knew the answer to the question that was still nagging me:

"Dick, who was *The Shadow*?" He was an older guy. I was sure he must know.

"Sorry, buddy, no idea."

Although we were still in the same prison, we gave this section its own name: The Desert Inn. Mike joked about all the single-star luxuries of The Desert Inn: the daily bathing, the three cigarettes a day doled out and lit by our generous guards, the wooden beds that were slightly less uncomfortable than concrete. It was still almost as cramped as The Heartbreak Hotel on the other side, still served up the same boring daily rice and soup and bread, still required us to bow to the enemy, and still forced us to crawl under our beds during the twice-daily air raids.

"But hey," Mike said, "at least the Desert Inn cleaning staff (namely us) does an excellent job of ensuring there are no rats, roaches, or spiders under the beds."

Mike developed deep, painful boils all over his body, abscesses really, until it got to the point that he could not lie down without excruciating pain. They were bad enough that a guard came in one day and led him out of the cell, and when Mike came back his skin was almost completely painted red with iodine or some other medicine. His abscesses seemed to get smaller after that, though they did not go away entirely.

Dick developed a boil at the corner of his eye. He constantly picked at it, though Mike and I insisted he was making it worse. What's more, I told him, "You're grossing me out."

I felt bad for my new friends, but couldn't help feeling relieved it wasn't me. Then I broke out in boils. They looked like warty red pickles. I had them on my navel, my shins, my heels, my hands, even one between my eyes. They weren't nearly as bad as Mike's, which ran deep into the muscle. Mine seemed less painful, though they definitely looked more disgusting. They rose to the surface, festered, popped, and drained pus. Nobody took me to the infirmary. That was okay with me. I didn't trust Vietnamese medicine anyway, and ever since my stay in The Rotten Cell, the sight of Mercurochrome or iodine or whatever it was made me queasy.

After three or four days of relishing the novelty of new friends, we more or less talked ourselves out. Boredom set in once again. We made up games based on TV game shows back home, like *Match Game* and *Password*. We made up a spelling game, and I soon realized how much more educated my roommates were than I. They knew the spelling and meaning of so many words. I tried not to think about that too much because it led me into a spiral of frustration, thinking about how I had volunteered for military service to escape college.

Back at East Aurora High, I had spent a lot of my young energy hanging out with friends, losing to them at nickel-ante poker, fighting over girls, and generally raising hell. I could never forget the day my best friend, Danny, and I were shooting at barn pigeons with our bows and arrows when we saw a bull in the barn with its back to us, and I tickled its balls with my bow. The bull chased us across a field. We launched ourselves over a fence just in time to avoid being gored. It was a terrifying, giddy rush, and we couldn't stop laughing. The point was, I was more into hanging out with Danny and my other buddies than doing homework, and my grades began to drop.

Poor grades were unacceptable for a first-born son, at least in my father's house. Senior year, Dad sent me to military school, hoping I would focus on my studies and make it to college, something he had never done. I grew apart from my buddies, whose lives back home in East Aurora went on without me. I was angry with my dad for a long time after that. But I had to admit, my grades did improve.

When the University of Toledo accepted me I was determined to please my dad, so I decided to study chemical engineering. He often talked about a friend of his who owned a successful engineering company, and I knew how much stock my father put into other people's images of success. In high school, I had nearly blown myself up during a chemistry experiment, but that didn't daunt me, and the idea of a challenge energized me. Not to mention, math was my strongest subject.

Turned out I hated chemical engineering.

Miserable at school, struggling to keep up, and lacking the college friendships I had imagined would make up for the loss of my high school buddies, I was a prime candidate for anybody offering a way out. On the other hand, failure was not an option in my book. So when the Navy showed up on campus my sophomore year and offered me a slot in the flight program even though I had only completed two years of college, I couldn't believe my good luck.

Even though they were older, sometimes Mike and Dick made me miss the friends I left behind in East Aurora: watching American Bandstand and drinking Coca Cola with David, or grabbing onto car bumpers on icy roads with Danny and letting unwitting drivers drag us on a wild ride—until the day a driver backed up and dumped us in a snow bank. Sometimes I looked at Mike and Dick and wondered what it might be like hanging out with a couple of college roommates right now instead of a couple of POWs.

Then on June 13, the guards opened the door to our cell and announced, "Roll up!" None of us knew where we were going, if we were going there together, or if we'd be separated. I realized these guys had become my friends.

The guards blindfolded us, walked us outside, and loaded us into a truck. The shuffling bare feet of other prisoners and the voices of the guards told me they were moving maybe half a dozen prisoners. We only drove for about five minutes before they ordered us out of the truck again. We walked into what felt like a large room, where a guard ordered us to put down our gear. Then we filed past a guard who held out a box of cigarettes. One by one we each took a cigarette, and one by one he lit them. I heard a door shut us in. An ear-piercing whistle startled me, followed by the loud whoosh and rattling sensation of a train rushing past us no more than thirty feet away. It felt as if the train were passing through the middle of the room.

We're at a train station, I thought. *Maybe they're putting us on a train to send us home. Maybe the war's over.*

10

ANTS

I was not standing inside a train station as I had hoped. The blindfolds came off, and Mike, Dick, and I found ourselves still together, but standing in a different cell in a different prison. The room was much larger than our last one, perhaps fifteen feet wide and thirty feet long with a twenty-foot ceiling. We paced the room for a moment in silence, stretching our arms and legs in the unexpected luxury of space. There were no stocks in sight. It was still far from comfortable: the pallets were hard wooden boards set on the floor, the windows were boarded up to prevent us from seeing out, and only three nearly invisible holes in the wall provided meager ventilation. We arrived at night, but the heat was already unbearable.

Come daytime, the heat intensified until the air was almost steaming. After breakfast, the guards sent us outside to do chores: pulling weeds, clearing trash, and cleaning empty rooms. We also made coal balls for the Vietnamese to use in their stoves to cook camp meals. The three of us sat outside around a two-foot pile of coal dust mixed with water, grabbing handfuls of muck and shaping it into lumps about the size of baseballs. We were to spend an hour every day patting those wet coal balls and lining them up on the ground.

The place was strangely quiet. We were three of only nine POWs in the entire ramshackle prison, which had clearly not been a prison before the war. We called it The Plantation, partly because of its resemblance to an actual plantation, and partly because our captors were using us as slave labor. Not that I saw our chores as a hardship. We laughed at the simple pleasure of moving our limbs and feeling the sunshine again on our pale skin. The guards did not punish us for laughing.

Despite rules against talking, we were able to figure out who our fellow inmates were. We snuck peeks through cracks in the door of our room and caught sight of the deep sunken eyes and big pointed nose of the famous Dick Stratton. That impressive honker earned him the nickname "The Beak." Stratton had made international news about two months before I was captured, when the North Vietnamese presented him at a news conference to prove that American prisoners were being treated well. The public relations stunt backfired. The North Vietnamese received bad press when Stratton acted as though he were brainwashed or drugged, repeatedly bowing in submission to his captors. Mike, Dick, and I agreed he still looked brainwashed. Within days we determined that wasn't it. His vacant look seemed to come naturally. Later he said he had purposely put on the puppet act at that conference in the hope it would heighten suspicions that North Vietnam was mistreating prisoners.

Stratton's roommate was Doug Hegdahl, a seaman apprentice who had served onboard a cruiser engaged in a duel with NVA shore batteries. He went topside during a nighttime bombardment, when a blast from the ship's five-inch gun mount jostled the deck and toppled him overboard. The North Vietnamese picked him up in a boat and took him to Hanoi. When prisoners made a competitive list of who had survived the highest and fastest ejection, we all gave Doug the award for surviving the lowest and slowest: fifteen feet at fifteen knots.

In those early days at The Plantation, we also spotted Larry Bell, Kay Russell, Dewey Smith, and Tom Hall.

The NVA maintained reminders throughout the prison that they were heroes and we were criminals. We passed a black-and-white photo exhibit of images from the war posted on the walls outside our rooms. One eight-by-ten of a nine-year-old boy on a bike had a caption that read something like *The Undaunted Messenger*. Other photos showed victorious North Vietnamese attacks on South Vietnam, or Vietnamese people running from U.S. air raids. One photo showed a disturbing image of a dead baby who had been killed by an American bomb.

In each prison we learned a routine. Although routine was boring, it gave me a sense of normalcy, which seemed worth clinging to so far from home. We woke at about six every morning, and the first thing we did was empty our five-gallon, black-painted, metal honey-buckets—*shit buckets* we called them. Back at our room, one of the guards lit a cigarette for each of us with a hot punk and then locked us back up to smoke and wait. Half an hour later, they let us out again to do yard work. Then we bathed. The Plantation had a large stall with several showerheads, but they didn't always work. Outside the stall sat a huge cistern. When the showers didn't work, we bathed at the cistern, filling buckets with water and washing ourselves and our clothes with bars of cheap soap.

After all that, it was still only about 10:00 a.m. Breakfast time. The guards let us out to collect our plates from a table in the courtyard, one room at a time. Our room was typically the last of the nine rooms in our section, so our food was usually cold. We went back to our cells to eat, back to the cistern to wash dishes, back to our rooms for another smoke, and back to bed to doze through the hottest part of the day.

At two or three in the afternoon, we went back outside for more chores, back inside for dinner, back outside to wash dishes, then back to our room to stare at each other or at the ceiling, to whisper stories or gossip about the other prisoners, to fantasize about being home with our families or invent games to waste away the evening until we fell asleep.

Then we'd start all over again.

One day a French film crew arrived to make a movie of us working. We were excited at the possibility our families might see it and know we were alive. All of us maneuvered toward the camera, trying to get our faces into as many shots as possible, hoping we wouldn't end up on the cutting room floor. The NVA just wanted the world to see how well they treated us.

That wasn't entirely a lie. At least the guards at The Plantation did not manacle us, cinch us, tie us, or slap us. Despite the occasional air raid, it seemed we were in for a peaceful summer.

After we cleaned up the new prison and made it more livable, more prisoners trickled in. By summer's end, The Plantation had about fifty prisoners. Mike, Dick, and I continued our surveillance through the cracks in our door to see who passed through the courtyard. That's how I spotted two pilots from my squadron: Lieutenant Junior Grade Read Mecleary and Lieutenant Commander Jim Pirie.

Read was hobbling on crutches and practically dragging his legs. It hurt just to look at him. I didn't know him well, but that didn't make me feel any less sorry for him. He had been shot down over Kep Airfield in North Vietnam, and his wingman told me years later that he had seen Read's plane list sideways and then simply disintegrate before his eyes. Read managed to eject, but when he landed, both legs were broken sideways at ninety-degree angles from each knee.

One day I peered through a crack in our door and saw a huge schnoz that did not belong to The Beak. I said, "There's only one person in this world who has that nose! That's Jim Pirie."

I had hit it off with Jim the moment we met in the squadron. We always had an easy time simply hanging out and laughing. He was one of those guys who seemed comfortable in his own skin, who didn't worry about what other people thought of him. I admired that.

Over time, despite the rules against talking, stories circulated throughout the prison: first through random whispers, then through stories passed to roommates who were later moved to other rooms, and later through more complicated code systems. It was via some combination of those shifting tides of communication that I learned Jim's survival story.

Our squadron used to make fun of Jim's shoes. He was our only pilot who kept the shoes the Navy had given him in pre-flight training: low-cut, suede, lace-up boots that we called *boondocks*. When the Navy selected us for the jet program, they gave us new smooth leather boots with metal toe protectors to keep our toes from getting mangled if we had to eject. The leather boot was a high-top with laces, but most of us paid a cobbler to put a zipper in the side so we didn't have to mess with the laces. I thought they looked sharper, sturdier, and safer than those little nothing boondocks, but Jim refused to wear the "fancy" boots, no matter how much we ribbed him about his "loafers," "sneakers," "girlie boots," or what-have-you.

"Kill your feet if you want. These boondocks are way more comfortable," he said.

On his final bomb run over Hanoi, Jim got the last laugh. His plane took a hit in the nose section, which punched the metal in, pinching his feet between the rudder pedals and instrument panel. Had he ejected while wearing the stiff, high-topped boots the rest of us wore, the smashed panel would have torn off both his feet. Instead, he simply slipped his feet out of those low-cut boondocks, pulled the eject handle, popped safely out of the cockpit, and landed in his socks.

Every morning, two of us walked past Jim's room to empty our shit bucket. One morning, when there was no guard nearby and the door to his room was open, I took a chance.

I slowed my gait, turned, and whispered, "How you doing?"

Jim gave me a thumbs-up.

"How's my wife doing?"

Another thumbs up.

"Am I listed missing or captured?"

"Captured," he said under his breath, just loud enough for me to hear.

Hearing that single word spoken by a true friend filled me with gratitude.

Although our captors cooled down the threat of punishment, the summer continued to crank up the heat. As Americans, we were used to thinking of cloudy, rainy days as bad weather. Now, the rainier the weather the better. We looked forward to thunderstorms, especially if they were windy or brought driving rain, anything to cool us down. As soon as each storm ended, the heat ramped up again, until it pressed on us as insistently as the roar of the cicadas.

When our cramped quarters passed 100 sopping degrees, the NVA was no longer our biggest enemy. We were. Not that Mike and Dick were unfair, greedy, or inconsiderate, but human nature and instinct took over. Lock two guys in a hot room and they'll form a team. Lock three guys in the same boiling room and sometimes two will gang up on one.

Little things loomed large.

Mike and I usually got along, but on the hottest, most boring days Dick and I got on each other's nerves. Some of that was probably our age difference. I was 23 and resented that I was wasting my youth behind prison walls. Dick was 35 and probably found my youthful impatience irritating. In my darkest moments, Dick's presence reminded me how much older I might be when I got out. In Dick's darkest moments, I probably reminded him his youth was over. We both missed our wives, but he had been married longer. He had two kids at home who might grow up without a father. I didn't have children. At any rate, although Dick and I liked each other, our different habits drove each other nuts. Unlike regular roommates, we had no option to move out or take a break from each other.

Dick talked a lot about his wife's cooking. He had a long list of things she made that were the best he ever ate, including her moist homemade cornbread.

"Big deal," I said. "Food is food." But I thought about Pat walking out of the kitchen carrying one of the only three specialties she knew: Baked Alaska piled high with meringue.

When our Vietnamese prison food came, Dick usually complained. One day we had pumpkin soup. That mushy orange glop reminded me of Halloween in East Aurora, when my buddies Danny, David, Jay and I stood on the railroad bridge over Main Street and dropped pumpkins onto passing cars below, hooting and hollering at the impressive splat, laughing at the sight of pumpkin guts, running for it when we cracked a windshield. The rest of the year we got caught breaking plenty of rules, but we never got caught on Halloween. To me, pumpkin soup smelled like youth, freedom, and possibility. It tasted like my whole life still in front of me.

Dick hated pumpkin soup. "This orange sludge is gross." Then he picked at his bread. "I can't eat this. It has bugs in it."

"It always has bugs in it," Mike said

I'd never been much of a diplomat, so I made a show of mixing the pumpkin soup with my rice, swirling it around with bread as if it were Mom's home cooking. "Mmm-mmm, delicious. Pumpkin soup with bread. Oh boy, my favorite!"

In reply, Dick farted.

I stopped with my food in midair. "Aw, man! Not while I'm eating."

"Typical Navy. You can dish it out, but you can't take it."

To ease boredom, we made playing cards from fifty-two sheets of rough manila toilet paper. We had fashioned a previous set of cards back at The Heartbreak Hotel, and the guards there hadn't seemed to mind, but here at The Plantation our new guards had confiscated them. I didn't understand why. It wasn't as if we had any money to gamble. We hid the new set, either in our rolled-up clothes or in the middle of our remaining stack of shit-paper.

We played a lot of Hearts. I had played as a kid, so what harm could there be? Dick had to remind us of the rules. He was an excellent player and beat us regularly until Mike and I caught on. Dick gave us a hard time, crowing over every victory. That's how I discovered that card games can turn three buddies into two-on-one.

One day Dick won several games, and Mike and I exchanged looks every time he gloated. Without a word, through the unspoken agreement that a couple of guys can make when they've spent so much time in close quarters that they can read each other, Mike and I teamed up to prevent Dick from scoring. We wouldn't let him win a single game. We grinned when he wasn't looking, but mostly we maintained our poker faces.

This went on for three days.

On day three, Dick threw his cards down. "This is chicken shit!"

Mike and I laughed, but I had to admit Dick had a point. At that moment, I could see a reason for taking cards from prisoners. Mike and I apologized and promised not to gang up on him again.

Maybe the biggest difference between Dick and me was that I believed my real life wouldn't start until after I paid my dues in the service and started flying for the airlines, while Dick was career military. Maybe he resented my attitude that the service was not my "real" life. I seem to recall him leaning back on his cot, pillowing his arms behind his head, and saying, "I have fourteen years in, just made major. I can sit on my butt here for the next couple of years and still make colonel. Hell, I can spend six years here and retire with full benefits. All this time up here counts toward my twenty."

Meanwhile, I only had a year and a half before my service was up, and I was horrified by the idea of wasting a moment more. "Six years? Shut up. Nobody wants to hear it."

Dick shook his head. "Don't be a wimp. I didn't say we'd be here six years. Hell, the way the war's going, we won't be here more than a year or two, tops."

But I didn't want to hear his optimism either. It all seemed too long. Why did he have to talk about it at all?

Mike tried to broker peace. "Give it a rest, you guys."

Deep down I knew that Dick was just putting up a strong front. It wasn't me he was trying to convince that his time here would be short or that it didn't matter. *Of course* it mattered. Every day he spent here was a day his kids would grow a little bigger, get a little smarter, and forget him a little more. Every day was another day without his wife. When his wedding anniversary came, or his kids' birthdays, Dick always grew quiet. On those days I laid off.

The NVA soon put in a camp-wide speaker system. One morning, the squawk box in our room woke us with a crackling shout,

"Wake up! All American prisoners, wake up and listen to the voice of the camp!" The male voice followed this with announcements in broken English, listing North Vietnamese victories and American losses in the war. What a way to start the day! From then on, "The Voice" quacked propaganda every morning. One day he announced our American generals were nervous because they had lost more than half their Thunderchief F-105s.

"That's the most ridiculous thing I ever heard," I whispered.

Mike tilted his head, calculating. "Bob, that might not be too far off. We only built three hundred of them, and the guy who got shot down before me was number one-fifty-six."

On the Fourth of July, guards led Mike, Dick, and me up to The Big House. That's what we called the small mansion that served as The Plantation's headquarters, where our keepers had their offices and sleeping quarters. We exchanged apprehensive glances, but said nothing. Talking was not allowed. We always kept track of the dates and we knew it was American Independence Day. This might be an excuse for leniency, or it might be an excuse to punish us after our peaceful hiatus. The guards walked us into a room where a man in an officer's uniform invited us to sit across from him. His hair was impeccably combed—no pith helmet—his khakis pressed, his gestures more exaggeratedly polite than we were accustomed to during interrogations.

"On occasion of American July Fourth, Independence Day, we will like to chat with prisoners." I immediately recognized "The Voice."

He offered us cigarettes, pineapple, and tea. He served the tea in demitasse cups an inch-and-a-half tall with gold inlay and a handle too small for our fingers, perched on tiny saucers. His questions indicated this was a goodwill visit: "How are you? How is your health? What do you think of the weather? What improvements would you like to see?" Although the words conveyed concern, The Voice's tone seemed sinister. Soon, I realized he was only putting on the sinister tone to overcompensate for effeminate behaviors he could not hide.

At first we kept our answers noncommittal. But he leaned forward with an earnest look of concern. My roommates and I exchanged shrugs as if to say, What the hell, take a chance.

"Do you get enough to eat?" he asked.

"Yes, thank you," Mike said, "but it might be nice to see a small change sometimes."

"What would you like to see?"

"In America, we have more meat," Dick said. "We don't get much meat here."

The Voice seemed to consider this. "We do not have much meat in this country."

Shoot for an easier target, I thought. "We would like to see more potatoes."

"The American planes have bombed the potato factory," he countered.

We tried another tack. We told him we would like more ventilation. He didn't say anything. It became clear to me that this was all just for show. We gave it one last shot, asking for sawhorses for our pallets, to raise our beds off the floor. He gave us an odd look, as if he could not see how that would help. We explained that sometimes we liked to sit on the beds, and that there were a lot of bugs and spiders on the floor. He nodded but didn't respond.

The talk lasted about fifteen minutes. Then the guards returned us to our room and escorted the guys next door to their own meeting with The Voice. Prisoners continued walking past all morning. Mike laughed it off, saying, "Do we all feel better now after our little chat?"

Actually, it wasn't a bad day as days in prison go. The guards brought a special meal. The guys and I were surprised that our enemies would encourage us to celebrate American independence. The most memorable treat was the orange liqueur served in little cognac glasses, which we saved for dessert. We sipped as slowly as we could, marveling at how tasty it was.

Mike told us about a cartoon he had seen in *Playboy* a few years earlier. In it, two hobos sat in a junkyard next to a couple of beat-up cars. One held a half-empty bottle of wine with half the label missing. "So the hobo holding the wine says, 'Not a great wine, but a good wine!'" We raised our glasses, laughing, and drank to that.

We were surprised when The Voice later complied with one of our requests. The guards brought sawhorses for our beds. It was a mixed blessing. One morning, Mike leapt from his bed with a yelp, frantically swiping at himself. A scouting party of black ants had crawled into bed with him. He traced them to their source: crawling through a hole in the floor and up the leg of a sawhorse onto his pallet. He decided to put a stop to them by pouring water down the hole. Bad move. Water filled the hole, floating thousands of ants up and out. In no time, his bed was swarming. They got into all of his stuff, keeping him hopping for an hour, shaking out his bed mat, clothes, soap, toothbrush, enameled cup; you name it.

This time, it was Dick and I who shared a big laugh at Mike's expense.

Eventually the ants retreated. Mike moved the sawhorses.

That wasn't our last run-in with ants.

The Vietnamese tapped us to wash dishes for the whole camp. We didn't complain. It was a chance to get out of our room. We gathered up the metal dishes left outside each room and took them to a small hut with a thatched roof and concrete floor dominated by a large double cistern with a faucet. As soon as we ran the faucet, water trickled down the pipes and found its way into a drainage hole in the floor. This time, thousands of fat, rust-red fire ants flooded the room. Dozens crawled up our legs and bit us like mad. Fire-ant bites hurt worse than bee stings, so it was almost as bad as busting open a beehive. Amid a flurry of yelling and jumping, Mike grabbed a bucket, filled it with water, and threw it on our legs.

We figured out the only way we could wash dishes was to post two people on washing-duty and one on bucket-duty, constantly throwing water on us to keep the ants at bay. What a circus! I didn't know whether to laugh or cry. We had to do that twice a day for the whole summer, and despite our best efforts we were regularly bitten.

Ants weren't the only insects that shared our tropical prison, only the most organized.

One night, we lay under our mosquito nets watching a huge praying mantis fly around our room. It settled on the light bulb that hung from a long cord in the middle of the high ceiling, waiting for the flying meals attracted to the bulb after dark. Sometime in the night, the light went out. A while later a faint thrashing sound woke me. It was unsettling not knowing what it was, but I had learned to sleep through almost anything. I didn't see what had happened until sunrise. Ants had formed a procession six bodies wide, marching up the wall, along the ceiling, to the cord, and down to the bulb. They were carrying the mantis out of our cell piece by piece.

Mike said, "Holy shit, that's what I call efficient."

It reminded me of the night I was captured, when I saw the unending lines of Vietnamese supply trucks winding through the dark, carrying supplies to the South. The praying mantis might be big, but it was outnumbered. The ants could not be stopped. I kept that thought to myself.

One night, I was balanced on the communal shit bucket when an air raid came without warning—it happened sometimes. *Blam!* The ground shook with the ferocity of an earthquake. A moment later, the air-raid siren sounded and the lights went out. Mike and Dick dove under their pallets, but I was in the middle of a bowel movement. Even once I finished, the prison lights were out and everything was pitch black. I felt around the concrete floor to try to find the shit-paper. No luck. I was screwed. Jet engines roared, anti-aircraft artillery banged away, bombs fell maybe less than a mile away. Over all that, I could hear my two roommates laughing their asses off under the beds.

Five minutes later the planes left, the sirens stopped, and the lights came back on.

I wiped myself, Dick and Mike crawled out from under their beds, and we all nearly cried with laughter. We fell asleep chuckling about their heroic lack of effort to save me, and how close I came to earning a Purple Heart and having to explain it to the folks back home.

It was good to remember that my roommates weren't the enemy.

As for the guards, even though they were the enemy, mostly they were just a fact of life. They were captors and we were captives, and we played our roles. One guard started out a pretty friendly guy, smiling and joking and going easy on the rules, but after a couple of months he turned sour, slamming doors and making us wait an hour before he'd give us a cigarette when he used to do it right away.

"What crawled up his butt?" I said.

"Maybe a prisoner gave him a hard time and he's taking it out on all of us," Dick said.

But Mike's comment was, "Familiarity breeds contempt."

There was another guard we called Bucky Beaver because he had an overbite. Bucky guarded us while we showered, and he had a habit of staring, so one time I stared right back until he grew irritated—maybe even felt threatened—and pointed his rifle at me.

I ducked behind the cistern, laughing nervously, "I didn't do anything. Don't shoot!"

"Quit bugging him, Bob!" Dick hissed. "You want to get us killed?"

I got on Dick's case for being a pussy, but I knew he was right.

11

THE BOY SCOUT

We called our section of The Plantation "The Warehouse," because it was a sort of warehouse for prisoners. The Warehouse quickly became part of a vast web of secret communications that spread throughout the entire camp—an imperfect web perhaps, with occasional gaps and breaks, but a web nonetheless. Every camp had one. I suppose it was a natural result of the needs of human nature combined with the habits of military structure: the drive to communicate.

At The Warehouse, the room I shared with Mike and Dick was Room 1. Within a few days, we established voice communication with Room 2. That was fairly straightforward. One of us would maintain lookout through the cracks in the door to see if guards were coming, while another would stand in the back corner and whisper through a crack in the wall to the prisoner in the next room. This was the easiest, most reliable way to relay information, such as the names of new prisoners, the stories of their capture, and the state of their health.

Most of the time voice communication wasn't possible, so we used a centuries-old prisoner code for tapping messages on the walls. The tap-code was simple, but it was ingenious in its efficacy. It was based on a square grid with twenty-five letters of the alphabet; for convenience, *C* doubled as *K*. We also shortened the word *interrogation* to the single letter *Q* for *quiz*. To tap out a particular letter, we would tap the column number of that letter, pause, and then tap the row number. Then we would pause again and go to the next letter. This method was surprisingly fast once we got used to it, much like Morse code. It was not as fast as normal conversation, but we had plenty of time to kill.

If we wanted to share information with prisoners on the other side of The Warehouse, prisoners who didn't share a wall with us or with our next-door neighbors, we waited for shower time. There were five shower stalls in a row that shared a common drain. The guards could not maintain a direct line of sight with all the stalls at once, so when a guard would leave, one of us would get down on hands and knees and talk through the drain to the guys on the other end.

The communication system was more about maintaining morale and camaraderie than anything else. We discussed a variety of things, but a typical question was along the lines of, "Hey, what's been happening down there the last couple of days?" That's how we kept up on information like who went to quiz, who got punished, the names of new shoot-downs, or the latest on the war from new prisoners. We put more confidence in war updates from new prisoners than in the propaganda The Voice fed us. Good news or bad, these communications made me feel as if I were still a member of American society, even if our part of that society was just a small band of prisoners cut off from the outside world.

On occasion, we needed a backup to our verbal and tapping comm systems. Sometimes the guards were especially vigilant and we couldn't make the slightest sound in the rooms or shower stalls without risk of discovery. In those situations, we wrote notes. We weren't allowed pencils or pens, but every so often the camp doc gave us little bottles of Mercurochrome or iodine for our boils, heat rash, jock itch, and such. We could then dip a spoon or stick into the medicine bottle, like an old quill-pen and ink, and scribble simple notes on our manila shit-paper. Sometimes guys mixed dirt and water with coal dust from our coal balls and wrote with that. The results looked like a note a first-grader might write in crayon, but for simple messages it got the point across, things like: "Kay Russell stopped puking," or "Jim Pirie switched rooms."

To make sure the intended party received the note, we would whisper as we walked by that person's room, "Note tomorrow." In the morning, we would drop the note in a prearranged area that we had chosen during previous communications. We usually passed notes when two guys filed out of each cell, room by room, to dump our shit buckets. While one of us dumped the contents into the latrine, the other would stuff a note into one of the cracks in the wall or ceiling. Then when the room expecting the note went to dump their bucket they could pick it up.

Breaking rules with my fellow prisoners was a far cry from the pranks I used to pull with my buddies in East Aurora only a few years earlier. Back then, I didn't care about getting caught. Back then, the stakes weren't so high. I remember the Christmas vacation when I came home from military school and a bunch of football buddies and I drove to the nearby town of Orchard Park to throw rocks at glass business marquees, just for the heck of it. We drove on to another town to spread more holiday cheer, and then—dumb kids that we were—we drove right back through Orchard Park on our way home. The police caught us with a bunch of rocks in the car. Luckily, the father of one of my friends hired a lawyer who convinced a judge that we were good kids who did a bad thing. We paid restitution and the charges were dropped.

If my inmates and I were caught breaking rules in a POW camp, we would have no lawyers. The Vietnamese would not shake their heads and say, "Boys will be boys." Our motives for breaking the rules were different, too. We weren't passing notes in the name of fun. This was a grown-up game. At the very least, we were playing to stay connected and stay sane. At best, these lines of communication were still serving our country, and might even help save lives: ours, our fellow prisoners', or those still out there fighting the war. If the NVA knew about the notes, the talking, or the tapping, they weren't giving us any indication.

During the final days of August 1967, when our room was so hot I wondered if one of us might die of heat stroke, two things happened. First, despite the heat, most of my boils went away. I never did figure out what caused the eruption. Second: the guards opened the door and said we were moving. I never figured that out either—why they moved us around so often. Maybe it was to prevent us from driving each other crazy and fighting, or from forming alliances that could threaten their control. Or maybe they had discovered our camp-wide communication system and hoped to disrupt it.

For whatever reason, Mike and I were moved to Room 13. This time they separated us from Dick. He moved down the row to a cell with four roommates. Mike and I moved into smaller digs with Lieutenant Junior Grade Larry Bell.

I had never seen Larry in action as a pilot, but I knew he had been shot down on his first cruise. He was only two years older than me, but where I was a compact five foot six, Larry stood about five ten, weighed 160 pounds, and sported six-pack abs. The solid abs must have come from his days on the rowing team at Annapolis. I was surprised he didn't move as gracefully as his athletic appearance suggested; when he did chores his movements seemed almost robotic to me. Still, between his chiseled physique and red hair, Larry had the air of an all-American Boy Scout. Sure enough, he was a Midwestern boy who had belonged to Junior Achievement in high school, sold *Grit* ("America's Greatest Family

Newspaper"), and graduated from the Naval Academy. He silently said grace before meals, never swore, and believed that as prisoners we must maintain our patriotic duty to serve our country.

What can I say? I admired the guy. We hit it off right away—even though he didn't know who *The Shadow* was either. I began to think of Larry as "The Boy Scout."

The Boy Scout's optimism was relentless, but at first his attitude inspired me. He congratulated us on what a great communication system we had and assured us that by working together we could make it even better. He shared his belief that it was part of our duty to stay fit and healthy, to exercise when possible, to eat even if the food was lousy, to be loyal to each other at all times, and to *never* befriend or give in to the enemy.

The Boy Scout seemed always to be on a mission to keep his mind and body active. He asked fellow prisoners for every scrap of information he could get, then did his best to memorize it all: quotations, poems, a list of all the U.S. presidents, definitions of words tapped on the walls by a fellow prisoner who had gotten his hands on a dictionary, and Spanish words tapped to us by a neighbor who knew the language. He often repeated variations on the theme, "I don't want my mind to waste away while I'm in here."

Sometimes the guards punished us for rule infractions, such as talking too loudly, passing notes, or not treating guards with the required respect. The punishments were minor. Usually they simply made us kneel on our concrete floors. A few hours of that could turn bones into pure conductors of pain, but the guards were not willing to stand over us for hours to make sure we stayed in position, so we rarely endured more than a fraction of the allotted time on our knees. Larry treated this as a team effort. If Mike or I got punished, he assured us, "We'll get through this together!" When the guards left I would stand up and move around to stretch my legs and back and give my knees a rest, while Larry watched out the cracks in the door for the guards. He would whisper a warning when he saw them coming back, so I could return to my knees. Mike and I did the same for him. I appreciated The Boy Scout's sense of humor. In the fall of 1967, The Voice announced propaganda about the battle of Khe Sahn almost every day. The American casualty numbers he announced were off the charts. At one point, he claimed the North Vietnamese had killed two hundred fifty thousand American troops. Larry said something like, "Well that's it for us then! We only had five hundred thousand troops in South Vietnam to begin with. Guess we're losing." We all shared a laugh over that one. I liked that Larry was smart enough to see through the bullshit.

One day we each received a large orange with our morning meal. It was the first orange I had seen since I'd been shot down. I decided to save mine for a post-siesta snack. I put it into my metal cup, covered it with the lid, and placed the cup under my pallet. Then I went to sleep. A few hours later, I woke up, reached for my cup, and removed the lid. The orange was gone. I looked up to see both Mike and The Boy Scout staring at me.

I gave them a suspicious look. "Where's my orange?"

"When you didn't eat it, we thought you didn't want it," Larry said. "So, we ate it."

"You gotta be kidding me."

"Nope, not kidding," Mike said.

I could not believe it. Here I was in a communist prison camp, I had not seen an orange in more than six months, and my roommates ate mine! I felt betrayed by these guys who were supposed to be on my side. How could I trust them again?

"Man that's cold!" I pumped my fists, clenched my jaw, breathed hard. I was trying not to lose my shit.

Larry ended my suffering. He pulled the orange out of his kit and presented it to me. My mouth dropped open. Then I joined in the laughter.

We were getting along famously.

Not that Larry was all about having a laugh. He could also be serious. I accepted that part of him as well, for a while.

On Vietnamese Independence Day, we received another special meal, much as we did on American Independence Day. This time the extra treats included fresh pineapple, peanut candy, extra bread, and extra meat. The meat came in a stew that reminded me of dog food, but it was pretty tasty so who cared about appearances?

Larry did. He warned Mike and me that the meat was an obvious tactic to brainwash us into celebrating communism, or something along those lines.

"Maybe," I said. "But then what were they trying to brainwash us into when they gave us extra food on American Independence Day?"

Larry's gung-ho patriotism was steadfast. In his campaign to improve our prisoner communication system, he sought to improve our writing utensils, simplify our note drops, and clarify our code tapping. He took over writing all the notes for our room. He also became determined to establish communication with a few colonels who had checked into the camp over the summer. After three months, they still refused to communicate with any other prisoners. We didn't even know their names.

Before I met Larry, Dick and I had tried to make contact with a couple of the colonels. We were sweeping underneath the window of their room, and I whispered, "Prisoner, say your name." The room seemed to go very still and nobody inside said a word.

After that, I had tried to talk to the colonels at the showers, hoping the cover of running water might make them more comfortable.

That time, one of them at least spoke. "I'm not going to give you my name. I was shot down June eleventh."

I tried to put him at ease. "I don't know what kind of treatment you guys went through as colonels. Everybody we know here went through the wringer at first, but since we've been in this camp the worst punishment anybody's had so far is kneeling for a couple of hours."

Still he didn't budge.

I wasn't the only one who tried. The colonels refused to talk to anyone—at all. They were so silent we couldn't determine the reason behind their silence. I assumed their initial interrogations had scared them shitless. Maybe as senior officers they knew more sensitive information and therefore believed they had more to fear from the enemy. Maybe they were right.

Larry insisted we give it another go, and Mike and I agreed to try. We finally talked to a prisoner who had been the back-seater in the June Eleventh Colonel's jet. That's how we found out his name was Colonel Herve Stockman. Not long after that, my roommates and I were bathing at the cistern when Colonel Stockman and another Air Force colonel were in the stall next to us. Larry and Mike watched for guards while I leaned over and talked to the colonels through the gap under the concrete divider that separated the stalls.

"Colonel Stockman, we've established a communication system, but we need a leader," I said. "We think you and your roommate are the two most senior prisoners in this camp. What do you say?"

"My roommate does not want to play ball."

"Your roommate?"

"Colonel Jim Hughes." That was still Stockman talking.

"Tell him about our comm system," Larry whispered to me from his post by the door.

I explained about our tap code and our note drops.

"Sounds like you have a pretty shoddy operation," Stockman said.

I thought, You've gotta be kidding me! Here we were, working hard and taking risks to set up a communication system while this guy provided no leadership for two months, and now he was criticizing us? "Colonel, we're doing the best we can with what we've got. It's not like having a telephone, but it works and we do get some information through."

"Excuse me, but I have to finish washing my clothes."

The Boy Scout could no longer contain himself. To my recollection, he charged over to the wall, dropped to his knees, and hissed into the gap, "Fuck you, Colonel!"

Mike shrugged, but Larry's response felt correct to me under the circumstances.

I don't know if Larry's persistence convinced Stockman, or if Stockman came to realize on his own that we were in this for the long haul and he would need friends to get by, but the colonel did ultimately come around and join our comm system. We discovered that he was indeed the senior officer at The Plantation. Larry seemed to feel vindicated, but I felt disappointed. Knowing that our first-in-command was the most reluctant to get on board with leading us through this trying experience added another crack in my faith that we could rely on our leaders.

I was leery of Stockman's leadership from the get-go, thanks in part to Larry's if-you're-not-with-us-you're-against-us attitude. It took longer for me to see the cracks in my image of Larry Bell. Sometimes I still envied his optimism, but I began to think his ability to see the bright side of everything might also be a reflection of his ignorance about the darker side of life. Like Dick, he kept talking about how soon we were going to get out of there. Unlike Dick, I didn't think he said that from hope but from an apparent inability, maybe even a refusal, to see the reality of our situation.

I'll admit, at first I was almost as reluctant as Larry to entertain the notion that America might lose the war. In that sense, we POWs all had a bit of Larry Bell in us. We had never seen America lose, so how could we imagine such a thing? When Larry railed against communism as a threat to world stability and freedom, I still agreed. But I wished he would give it a rest.

After a couple of months, his patriotic zeal got on my nerves so much I began to question whether I agreed with him about communism being evil. I agreed it was a bad idea but no longer felt so sure it would ruin the planet. I began to consider the danger of blind faith in, or blind hatred of, a single idea, any idea.

Larry was idealistic. Guys like him believed that, because a war had landed us here, we still had a mission as warriors. The goal of this supposed mission was unclear: we had to decide that for ourselves under a haphazard collection of leaders who just happened to have the bad luck to end up here with us. Whatever the goal, backing down was not an option for Larry.

I began to see myself as more practical than him. My interrogations had taught me the power of pain and fear to turn a loyal man into an empty puppet. I still agreed with Larry that we needed to guard our lips from making political confessions to the enemy, that we should not give the North Vietnamese fuel for their propaganda machine. But I didn't worry as much as Larry did that everything I confessed under duress would hurt my country. People expected the Vietnamese to trash America. Anything I said under interrogation, whether the NVA reported it correctly or not, would be a drop in the bucket. My practical side told me that, in general, the less time I spent kneeling on concrete the better. My most important goal as a prisoner became surviving so I could go on with my life.

At first, I said nothing to Larry about the differences between us. I believed that if I disagreed with him he might label me a traitor. Over time, I came to believe that Larry's potential judgments of me could be more dangerous to me than my interrogators' questions or demands. I knew that many POWs had come home from the Korean War tarred for life by words their fellow soldiers passed around: "He's a traitor." It didn't matter that those guys had been tortured and that nobody could hold out forever under torture.

The possibility of being tortured by communists or labeled a traitor by The Boy Scout didn't bother me half as much as having no place to escape from even the most minor annoyances. One day a guard decided to remove the rungs of a metal ladder with a hammer and chisel directly outside our room. It took him two weeks to finish the job. Every day for two weeks he just hammered away, metal on metal. The noise threatened to drive me insane. Still, much as I hated my lack of control over my environment, I accepted that there was little I could do about it.

Larry, on the other hand, struck me as a man obsessed with controlling everything he could, especially our comm system. If he had something to communicate to someone else in the system, it was as if it needed to happen then and there. It seemed he was forever pressed against the door, looking through the cracks at other prisoners, trying to ID them so we could match the names we heard through the grapevine with the faces that passed through the courtyard.

His attitude came across as, "C'mon, Wideman! This has to be teamwork, man. Two sets of eyes are better than one."

My attitude was, "Relax. We can find out when we hit the showers."

I believed that Larry considered my attitude laziness, complacency, even disloyalty. "Where's your can-do attitude, soldier?"

"Look, we both know there's not much to be gained by looking at some guy a hundred yards away through a tiny crack in the door. Shit, I'd seen you walking by before, but when you moved in I still didn't recognize you up close. You could've been anybody. Right, Mike?"

Mike shrugged. "More or less."

The Boy Scout suggested I was making excuses.

One day I got so sick of his persistence, I said something like, "Larry, the comm system works fine. Everyone's okay with it but you. There's no crisis here. What's the big deal?"

Larry seemed to think the big deal was that we didn't care enough. Someone who listened to him might have thought we never would have set up a comm system without him. Mike pointed out that we had communication in the last camp before Larry arrived. But it was Larry who helped convince Colonel Stockman to get on board, it was Larry who warned us never to let our guard down, and in my opinion it was Larry whose self-righteousness filled the space between us—as if he believed Mike and I would have sold out to the enemy if not for him.

He made such a public display of praying silently before meals that Mike and I couldn't comfortably eat until he finished.

I asked him about the Bible passage that says: "When thou prayest, enter thy closet, and when thou hast shut thy door, pray to thy Father which is in secret..." (Matthew 6:6)

He countered with his own passage: "Neither do men light a candle and put it under a bushel, but on a candlestick; and it giveth light unto all that are in the house." (Matthew 5:15)

He didn't pause to consider my ideas before knocking them down. It seemed I was not as God-fearing, dedicated, or patriotic as he was.

When he had exhausted every idea he could conjure to whip Mike and me into a communication frenzy, he would find other outlets for his Boy Scout energy. He used leaves and shit-paper to make himself a hat and a little pouch to carry his things around. His need to go above and beyond combined with his ingenuity would have been ideal in a normal environment. During peace-time, he would have had a better chance to impress his superiors, and might have been less inclined to pester his peers. He would have been quickly promoted up the chain of command, or might have gone into the private sector to become an entrepreneur or corporate exec. Larry's drive would have led him to change the world.

We were not in a normal environment, however. We were stuck in a cell with the same people day in and day out, waking, sleeping, eating, pissing, working, showering; together, always together, under the glare of lights that never turned off. In that situation, Larry's can-do attitude sometimes made me want to smash his face in, whether he deserved it or not.

Even after my admiration for Larry Bell vanished, I still envied him, not for his rippling abs or blissful ignorance, but for the seemingly charmed life he had lived up until his plane crashed. Larry talked a great deal about his family, and when he spoke of his father it became clear the two were buddies. His dad did those things I thought dads only did in the movies: played catch with his son, helped him with homework, showed up for every achievement, even supported Larry in goals that differed from what his father expected. Larry told us that his old man was a loving husband and father who had never struck his wife.

For years my father had knocked my mom and brother around. He had spared me, the first-born in whom he placed his expectations, hopes, and dreams. None of the kids I grew up with ever talked about what happened in their families, and I never talked about mine, so I assumed my childhood was normal. From the sound of it, Larry's dad was the kind of dad every kid wished for. His father had given him the kind of support that made him truly believe everything would be okay. That was a foreign idea for me. Our differences ran deep.

The more time passed, the more I realized that my fantasy about bonding with fellow prisoners had been just that. In those early weeks alone, what I had missed was not so much *living* with my fellow men as *communicating* with them. I began to wish I could live alone again, provided I still had a way of talking with the others—if only by tapping on walls, writing iodine notes on shitpaper, or scratching messages on the bottom of cups and plates. More than I needed to connect with other humans, I yearned for one thing so many Americans prize above all, though I didn't truly feel its importance until it was gone: privacy.

As much as I wanted to get away from the North Vietnamese, my desire to escape The Boy Scout was stronger. The Vietnamese would walk out the door and leave me alone, but my only escape from The Boy Scout was sleep. Unless of course I fell asleep when he thought I should be paying attention. Then there was no escape at all.

12

HOLIDAYS

About two weeks before Thanksgiving 1967, a couple of Vietnamese men on bicycles rolled into camp. Each bike was loaded with live turkeys dangling upside down. The men had tied up each turkey's legs, roped the turkeys together in pairs, and then slung them, two-by-two, over the handlebars and fenders of the bikes. What a sight! That's how at least three-dozen turkeys rolled into The Plantation that day.

Dozens of human eyes peered through the cracks in nearby cell doors or cast glances while on the way to chores or showers, as the guards took the turkeys off the bicycles and untied their legs. It was the best entertainment we'd had in weeks, watching those black-feathered birds with bright red heads spend the rest of the day staggering around the courtyard trying to straighten out their wobbly legs. By evening they were strutting around like they owned the place.

The next morning, Mike and I passed through the yard to empty our shit bucket, and the turkeys were gone. I looked around, and finally found them perched in the trees. Several of them stood balanced on one leg with a head under a wing. I never did see them fly up there. It seemed as if they had vanished and reappeared by magic. Sometimes it was hard to accept that the world outside our cell continued to exist while we were locked away. Maybe that was why The Boy Scout was so fixated on staring out the cracks in the door.

The day before Thanksgiving the birds disappeared again. The next time we saw them they were on our plates for another special meal. It seemed as if I were marking my time by those holiday meals, those few days when my belly felt full and there was a bit of flavor or color on my plate to break up the boredom of prison life. I wondered if the Vietnamese chefs who cooked the turkeys felt confused by the American enthusiasm for eating such an unlikely bird.

The chunks of bird meat reminded me of hunting with my dad, though we usually hunted duck, not turkey. I remembered the pride I felt in hunting for our own food, and the feeling that my dad was proud of me.

I imagined arriving home from the war to find my mom standing at the stove stuffing a turkey. I would sneak up behind her and throw my arms around her in a bear hug. I pictured my father, mother, brother, and me sitting around a Thanksgiving table, holding hands and praying over Mom's cooking. Even if I were back home, it would have been a fantasy: Dad didn't believe in prayer, and my parents no longer shared Thanksgiving. They had divorced four years earlier.

Thanksgiving was one time I did not resent The Boy Scout praying, and I was grateful that he asked Mike and me to join him. No doubt we were all thinking about our moms, though none of us said a word about them. The turkey was tender and tasty. For a few minutes the only sound was three men chewing, sitting alone and silent on our separate bunks.

A couple of days before Christmas, Mike and Larry went out at about 6:00 a.m. to empty our shit bucket. It was still dark and I was snoozing on my cot when I heard someone say, "Get up!" I opened my eyes and was surprised to see the face of the guard we called Vada. He was a green recruit who had joined the staff a few months earlier, and he often made camp rounds with food, cigarettes, or the hot-tea cart. This time he was holding a chessboard and bag of chessmen in one hand, and a deck of playing cards in the other—not the makeshift games we cobbled together from sticks and toilet paper, but honest-to-goodness three-dimensional games. That wasn't what surprised me, though. What surprised me were those two words: "Get up!" Like most newbies, Vada had never spoken a word of English to us.

I scrambled to my feet and bowed, my brain racing, trying to remember all the times my roommates and I had spoken to him in English, sometimes making fun of him with a poker face, getting a kick out of his lack of reaction to our taunts. We had bowed deeply while saying things like, "Good morning, you stupid bastard." Had he understood what we said? Would he retaliate?

"Good morning," he said. "How are you today?"

"Fine, thank you," I said. "And you?"

"I am fine, thank you."

"I didn't know you spoke English."

"I have been learning." He sat on the cot across from me and gestured that I should do the same.

"Are you married?" he asked.

That old routine. I wondered if this was leading to reward or punishment. "Yes."

"How many children?"

"No children."

Then he threw in a new question, holding up the chess set. "Do you play chess?"

I perked up. "Yes."

"I have chess, checkers, and cards. Which would you like to play?"

"Maybe chess."

That was about the time Mike and Larry came back from emptying the bucket.

Vada gave all of us cigarettes, lit them, and chatted a moment longer. His English was limited. Still, Mike and Larry seemed to be working hard to maintain their cool, too dumbfounded to speak.

After he walked out and closed the door, Mike broke the stunned silence. "You know, that's like waking up one morning and your dog starts talking to you."

After that, we watched what we said in front of all the guards.

We got a kick out of playing with a real chess set—a small one, sure, but it had actual kings, bishops, and knights. A miniature war game. If the Vietnamese had considered the resemblance, maybe they wouldn't have offered it to us.

On Christmas Eve, the guards went through the prison, selected a few prisoners, and took them away. It took a few hours of tapping code from wall to wall to figure out where they had gone. The Vietnamese had taken the Americans in cars to see the holiday lights downtown.

"I didn't know they celebrated Christmas here," I said.

"Even back home, not everyone who celebrates Christmas is religious," said Mike.

The Boy Scout said something like, "But a bunch of communists? No way! They're probably looking at some sort of New Year's lights or something."

Later, the word tapped up and down the walls was that they got to see real Christmas lights: red, green, white, and multicolored, the whole bit. Those of us left behind still had one thing to be grateful for: no air raids on Christmas Eve. It truly was a peaceful night in Hanoi.

Early Christmas morning, our door stood open for a rare moment. Larry, Mike, and I stared across the camp at a fellow prisoner who was feeding turkeys while a Vietnamese with a movie camera filmed him. The prisoner was smiling as he fed the birds.

I thought I heard Larry mutter, "Why's he smiling? Doesn't he know they'll use that film for propaganda? He's aiding the enemy."

The prisoner picked up a turkey that had gotten away from the bunch and carried it back to the others.

Larry whispered something like, "Put the turkey down, man," but for our ears only.

I nodded, but said nothing. I wasn't as angry with the other prisoner as Larry appeared to be. I figured he was a captive who had to do what he was told or face punishment, and I couldn't see a reason for him to risk disobeying our captors over something so unlikely to endanger American forces. But I did feel that the photo shoot was deceitful on the part of the North Vietnamese, because it was likely to give the public the impression that they treated us like kings all year long, when usually we only ate two lousy meals a day and the turkeys were an exception for Christmas.

A short time later, we were invited to The Big House for another holiday chat with The Voice. This time we sipped sweet coffee from the little gold-filigreed demitasse cups. We ate cookies, little coffee and peanut candies, and candied orange and lemon peels.

He asked us the typical polite questions: Were we getting enough to eat? Did we like our chessboard? How was the weather? The last time we had sat down with The Voice like this, Dick had been with us. I was curious to see how The Boy Scout would react. Would he give surly answers? Refuse to answer at all? After all his dire warnings against appearing to cooperate with the enemy, I was surprised that he was as subdued and soft-spoken as Mike and I.

On our way out of the building, the guards gave each of us a little package, which we didn't open until we got back to the room. The packages simply held more cookies and more candy, and I was glad. Anything besides food would have been depressing, a reminder that nothing could make up for missing Christmas at home. And, had the packages held something other than food, The Boy Scout might have tried to guilt us into declining gifts from the enemy. But Larry never argued with food, thank God. I laid out each sweet morsel on my bed and savored them one by one, the way a kid might count and recount the loot from a Christmas stocking. It was just a handful of sweets, but all of us agreed it was tasty.

At dinner they fed us more than usual, including turkey, salad, potato soup, and a small bottle of beer each. I was surprised they had the capability to prepare such a rich meal. This was almost a taste of home.

After dinner, a guard came by our room and asked us, "Are you Protestant or Catholic?"

All three of us answered, "Protestant."

"We can take a few people to church tonight," the guard said. "Would one of you like to go to church with us?"

They were not inviting all three of us, just one. There wasn't enough room at the service for all the prisoners in camp. All three of us turned him down. I had thought Larry might say yes since he was a devout Christian. But I seem to recall that when the guard left Larry expressed his usual suspicion: that it was another ploy by the communists to convince the world that they were the good guys.

Maybe that was true. That night we heard two sets of prisoners singing from across the camp. Later we learned from other prisoners that there were two services in two locations, each with about fifty prisoners—each with a slew of photographers snapping photos.

Now and then the Vietnamese offered other prisoners little trips into the city too, a few guys here, a few guys there, to see the war museum, the cultural museum, or the Tet fireworks celebrating the lunar new year. The guards never invited me to any of those outings, and I never asked to go. I didn't care much. The only trip I wanted to make out of the prison was the trip home.

The day after Christmas, Vada came by our room again and took the chess set away.

For the duration of winter, we had new bath-time rules at The Plantation. The Voice was gone, but his replacement, another smooth talker we called The Lackey—because he was a by-the-book guy—announced over the loudspeaker that we were not to shower or pour water on ourselves when we took baths. Instead, they wanted us to use a wet cloth to clean our armpits and to clean our crotch. I thought that was a good idea during cold months. Although the temperature never dropped to freezing, it did drop into the low forties. It was good to know that, enemies or no, they did not want us to get sick if they could help it.

Now and then The Lackey called us up to The Big House individually for quiz. Those sessions didn't feel like interrogations so much as reminders that they were in control, and that they had as much as or more information than we did on the American war effort. Sometimes The Lackey loaned us history books about Vietnam, China, and Korea, or newspaper articles about the Vietnam War, all of them slanted toward a communist perspective of course.

Starting in late January 1968, The Lackey reported news to the camp about the siege at Khe Sanh. Some twenty thousand North Vietnamese used a massive barrage of firepower to pin down six thousand U.S. Marines just a few miles from the Laotian border. The strategic location was pivotal for control of the Ho Chi Minh Trail through Laos. On day two, the Vietnamese hit the main ammunition dump, detonating fifteen hundred tons of explosives. The siege dragged on for months. Even Larry said it sounded bad, though the only reports we heard or read were from Vietnamese sources. It would go down in history as the bloodiest battle in Vietnam, and during quiz with The Lackey, he showed us photos of the horrors taking place there. I remember a photo of a chaplain praying with troops for God to extract them from this nightmare. I was glad I was in prison and not at Khe Sanh. I could handle being shot at in an airplane or sitting out an air raid in a prison, but I knew that I would not be worth anything in a foxhole.

Ten days after the Khe Sanh siege began, the NVA launched an all-out offensive on more than a hundred towns and cities throughout South Vietnam, during the wee hours of the lunar new year, which the Vietnamese called Tet. The Lackey didn't call it the Tet Offensive, just announced that a big attack took place in South Vietnam and that the U.S. lost hundreds of thousands of troops. If that were true, it should have ended the war. According to the inflated numbers both The Voice and The Lackey had been reporting, the NVA had killed every last one of our troops in the South. My roommates and I scoffed at the outlandish propaganda.

"Didn't I tell you that you can't believe a thing they report?" Larry said.

Still, Mike and I exchanged uneasy looks. If we could not trust their numbers, that meant we were essentially living in an information blackout. We had no way to be certain what was truly going on. The U.S. could have launched a nuclear strike and we might not have known about it unless it landed right in Hanoi. My only brother could be among the dead already and I would not know for sure until I got out of this place. The Vietnamese numbers might be laughable, but during the twenty-six days of the Tet Offensive I found it difficult to laugh.

Surely The Boy Scout must have shared at least some tiny doubt about America winning this war, though he would never admit such a thing to me and he took every chance he could to prove he was still in this to win. During one of his quizzes, he snagged a picture of Ho Chi Minh, which he then placed above the latrine where we dumped our shit buckets every morning.

By that afternoon, a message came tapping at our wall, passed on from Colonel Jack Van Loan a few walls down. He wanted us to know he had seen that picture perched in the latrine and had personally removed it. B-a-d...i-d-e-a our next-door messenger tapped. At least one of our leaders demonstrated common sense. Larry let us know he was ticked that not all of his fellow prisoners had his guts, but Mike and I exchanged looks of relief. So far, The Plantation was not a terrible place to stay, as prison camps went, but whatever was going on in Khe Sanh or in the South with this Tet attack, it wasn't good. If those battles didn't go North Vietnam's way, we knew that our captors' attitude toward us captives could shift at any time.

The Tet Offensive and the Battle of Khe Sanh were still raging when The Lackey came to our room one night after dark and said, "Okay, prepare to move to a new camp." My nerve endings tingled with adrenaline as my cellmates and I rolled up in silence. Did the move have anything to do with the tide turning one way or another in the terrible battles raging north and south of us? No way to know. We waited for half an hour, listening to movement outside: people shouting, doors slamming, engines rumbling. Tires rolled to a stop just outside. Then our door opened to reveal a jeep in the courtyard just a few feet away.

We threw our stuff into the back of the jeep and climbed in after it. A guard blindfolded us and then sat down in back with us as we moved out. I could see a little over the top of my blindfold, so I saw when we passed through the gate and I was pretty sure we headed southwest from there. A few streets were lit, and I could see and hear thousands of bicycle wheels spinning and whirring past: men in uniform, old people, children, young guys with girls sitting on back, arms wrapped around their waists. It seemed the entire city was on two wheels, except for us, and every now and then another jeep, slowly buzzing through the motor-less crowd.

We crossed some railroad tracks, took a couple of turns, and after about half an hour and maybe five miles of driving, the jeep stopped. The guard took off our blindfolds and told us to grab our belongings and get out. The driver disappeared down a dark dirt road leading to a long brick wall while we waited with a lone guard at the road's end. Just us, some rampant green vegetation, and the vast night sky. We took the rare opportunity to crane our necks and admire the beautiful night. I had never seen so many stars, every last one perfectly clear. The moon was bright, but even its luminescence could not dim the stars.

Our guard reappeared and led us down the road to the red brick wall. We passed through a gate into a small courtyard, and then up some steps to a long, low, gray cinderblock building where a turnkey waited. Our guard left, and the turnkey led us to the open door of a whitewashed cell smaller than our room at The Plantation, maybe thirteen by thirteen. He closed the door. As always, a single hanging bulb remained turned on. Darkness was a luxury we could barely remember.

We put our rolls on the three beds, which were wood boards again, each propped atop a trio of concrete slabs. We turned in slow circles to take in our new surroundings. The beds sat close together, leaving half the room wide open. But we all stopped to stare at something else: a huge, glassless, steel-barred window. The shutters were open, letting in a cool breeze, something we had not felt in months in our windowless box at The Plantation. I felt relieved because when it came to survival in Vietnam, heat was the element that worried me most. Still, we did not sit.

A half-dozen Vietnamese, guards I think, stood in a circle outside the big window, talking, so I kept my voice to a whisper as

I named the fear I knew we all shared. "Do you think this is The Zoo?"

Mike and Larry shrugged, but their wary expressions told me they were wondering the same thing. We had no clues. We had heard only rumors back at The Plantation. Two of those rumors stuck with us: The Zoo featured both an algae-coated swimming pool that nobody swam in...and torture. We had not seen either yet, but something felt off. Tense.

Dick Stratton, of catatonic press conference fame, had been to The Zoo. Back at The Plantation, Jack Van Loan had talked to my old roommate Dick through a wall, and Dick had passed word on to us that the NVA had beaten Dick like a rug to persuade him to go to that conference at all. Jack said that when the North Vietnamese found out Dick had acted brainwashed on purpose, they put him in a dark room by himself for a very long time. As tired as I was of sleeping in the light, the idea of being left alone in the dark sounded worse.

My roommates and I did not mention Dick. Just stared out our window and waited.

The door opened, and there stood The Voice, who had disappeared from The Plantation a few months earlier. His smile struck me as sinister, but that was normal for him. His uniform was nicely pressed, also normal. He wore a pith helmet. That was new. Mike, Larry, and I bowed. It was always weird to see Larry bow.

"How are you?" The Voice asked.

We each mumbled a noncommittal, "Okay."

He adopted the tone of a counselor at summer camp. "In this camp you must obey the regulations very carefully because the camp commander is very strict, not like the other camp."

His words made me feel flashes of hot and cold, a fever of uncertainty. He left and we sank onto our beds, exchanging looks that communicated one thing:

Looks like we're in deep shit now.

13

THE ZOO

A man can only focus on his fears for so long before they either consume him or fade into the background. My roommates and I debated what The Voice meant when he said The Zoo's commander was stricter than The Plantation's commander, but we were unable to reach a conclusion that first night. It wasn't as if we could move if we didn't like it, so we let it go and settled in like we always did: moved our shit bucket into the corner, selected our beds, put up our mosquito nets, and arranged our scant personal items.

Even a man with next to nothing personalizes his territory. I liked my things in piles where they were easy to grab when needed. Larry preferred his belongings folded and stacked to regulation perfection. Mike was a minimalist—as if this was a motel and he didn't intend to stay long.

Every so often a guard walked by, lifted the flap on our metal door, gave us a stern once-over, said nothing, closed the flap, and continued on his way. We agreed the atmosphere was spooky, but as we drifted off to sleep we relished the breeze drifting through the window.

The next morning, I woke to a sight I had forgotten: the warm reds and oranges of sunshine sifting through my eyelids. I opened my eyes to a brilliant gold sun shining through our open window. I had not seen a sunrise in more than a year.

"This is great," Larry said. Finally, something we could agree on. Breezes still stirred the air, so we took down our mosquito nets. We only had one bucket, and Mike volunteered to dump it. He reported back on his brief trip outside, "It's kind of dirty out there, not as clean as the other place." The guard had told him to dump the bucket right next to the showers. At The Plantation, showers and latrines were on opposite ends of camp.

Our meal showed up around ten o'clock, and we were surprised when we were among the first in line. Our soup, vegetables, and rice were hot.

After breakfast, we found out that the guys next door had made a crack in the wall between our rooms, so we could talk to our neighbors more directly than in any room thus far. One of the men identified himself as Navy Lieutenant Junior Grade Everett Alvarez Jr. I almost fell over with disbelief. Alfie was the first U.S. aviator taken captive in the Vietnam War, shot down during America's first air raid on North Vietnam.

Our neighbors verified that we were indeed prisoners of the infamous Zoo. So when our conversation grew lengthy I became nervous about the guards catching us and doling out severe punishment. But the guys next door didn't seem concerned. They said our building was isolated and wasn't checked as frequently as the others. The room I shared with Larry and Mike was one of nine in a section the prisoners called "The Office" because it looked like a small office building. The down side to our isolated location was that we had little communication with the rest of the camp.

Our neighbors clued us in that The Zoo's reputation was as much legend as fact, though they confirmed it was indeed a tough camp where torture was used to persuade prisoners to peddle Vietnamese propaganda. They warned us which guards were harshest, particularly one interrogator called "The Colt." We later nicknamed him "Dumb-Dumb" when we concluded he beat prisoners to compensate for his lack of intellect.

After we heard the lowdown from our new neighbors, I lay in bed and did the math on Alfie. He had been shot down in August of 1964. It was now March of 1968, so he had been a POW for three and a half years. I had been in for one year, and already it felt like status quo, like being a prisoner was now my life. But three and a half years? That was a sobering thought.

Nothing much happened for a few days until we each got our call for quiz with Dumb-Dumb. He wanted each of us to write yet another autobiography, something we had done over and over. In one of his typical fits of patriotic defiance, The Boy Scout finished up his session by stealing a pen from Dumb-Dumb's desk. When he returned to our room he wagged it in front of Mike and me with a grin. Despite my irritation, I said nothing. I had wearied of arguing with him.

About twenty minutes later we heard footsteps, and Larry stuffed the pen in our shit bucket. The door opened, and Dumb-Dumb was standing in front of us, his previously dull expression nearly rendered intelligent by fury.

"Did any of you take my pen?"

I said no, Mike said no, The Boy Scout said no.

"Whoever has taken my pen, give my pen back!"

"We don't have your pen." The Boy Scout's indignant defiance was impressive.

Even Dumb-Dumb could do math. He focused on The Boy Scout. "You were the last one. My pen was not missing before. Where's my pen?"

"I didn't take it," The Boy Scout repeated.

Dumb-Dumb shoved him to the floor. "On your knees, prisoner!" Then he stalked out.

Mike and I took turns standing by the window to tell Larry when it was clear so he could give his knees a rest.

Bill Baugh, one of the guys next door, whispered, "Where'd he hide it?"

I stepped to the back corner and whispered through the crack. "In the shit bucket."

"Well, they will look in the shit bucket," he said.

We searched for another hiding place. The room was bare except for our meager clothes, dishes, and toiletries. We looked up at the twelve-foot ceiling. "There!" Larry pointed at a two-foot-square hole covered with wire mesh, which led to a crawl space. There wasn't any light up there, so it seemed a perfect hiding spot. Almost.

"How're we gonna reach it?" I asked.

Larry's exasperated look told me I lacked vision. While Mike watched the window, Larry tossed the pen upward. It slammed into the ceiling, and he dodged it just before it clattered onto the floor. Several tries later the pen sailed between the wires and lodged atop the grating. Larry looked triumphant. I gave him a smirk to tell him he was kidding himself, but he ignored me.

Dumb-Dumb returned but said nothing to Larry, who had returned to his knees. Instead he ordered me to join Larry on my knees. Then he turned to Mike. "You're moving. Roll up."

Just like that, Mike was gone.

Larry said something like, "You know they're going to try to pit us against each other."

No kidding, I thought, but said nothing.

"We can't let them divide us."

I had little choice but to have his back. It was just the two of us now—almost. For the next twenty-four hours, our neighbors cleared for us. A couple of times Dumb-Dumb came in, slapped us, asked where his pen was, slapped us again when we played dumb, and left.

Sometime during the second afternoon, our door slammed open and three guards came in with a ladder. One of them hauled us to an empty room nearby and locked us in there. About thirty minutes later, he returned and led us back to our room. The moment they left, both our chins tilted up simultaneously. The barbed wire looked bent. We couldn't see the pen anymore.

"It's gone," Larry said.

My God, I thought. They wrapped Mike up and he told them where the pen was. Now they've got proof. I didn't voice my suspicion. I didn't want The Boy Scout to blame Mike.

Larry and I kept looking at each other, alternately kneeling and pacing depending on whether Alfie and our other neighbors tapped a warning that the guards were coming or whispered, "Clear!" A couple of times Dumb-Dumb returned to slap us again. The next day the punishment stopped, and he never said another word about his pen.

A few days later, word made its way back to us from Mike. He had not been questioned about the pen at all. Apparently, it was just a coincidence that the guys with the ladder had chosen that time to do routine maintenance. We never did find out what happened to the pen.

About a week after Mike got moved out, our buddies next door were moved too. For a couple of days, Larry and I were stuck alone together. Without Mike as a buffer, the bad blood between us festered.

On March 15, 1968, Larry and I upped the ante in our ongoing battle of wills. He repeated one variation or another on his favorite refrain: "I'm telling you, the war is almost over. America has North Vietnam on its knees."

"No way, man. We're not leaving anytime soon. Vietnam will end exactly like Korea. We'll cut 'em off until there's no live communists south of the 17th Parallel, and that'll be the end of it." In fact, cutting them off at the 17th Parallel had been our objective when I was shot down. "You watch. This thing's gonna be a stalemate."

I hadn't stopped believing in America. I felt pride in my country's persistent refusal to give up. It was a value my father had drummed into me: "Don't be a quitter, son. There's nothing worse than a quitter." But I had seen and heard enough since I'd fallen from the sky to tell me the Vietnamese weren't quitters either. I could see these determined people continuing their guerrilla war as long as there was one guy left standing in South Vietnam with a grenade.

The Boy Scout made me so mad, I couldn't think straight. I said that if he was so sure of himself, he should put his money where his mouth was. I told him I was willing to pay him a dollar for every day we were in Vietnam up to Election Day, November 4, 1968—more than seven months away—but for every day we stayed after November 4, he would owe me a dollar. That's how sure I was this war would go on and on: I was willing to start more than two hundred dollars in the hole.

Having The Boy Scout as a cellmate made the prospect of winning my bet pretty grim. If I was stuck with Larry for a while, a long war was going to feel like an eternity. So it was that I felt deep gratitude for the arrival of our new roommate: a son of a bitch and a patriot whom I could truly admire.

At more than six feet, our new roommate towered like Hercules over the two guards who brought him to us. His size alone wasn't the only way he filled the room; he was one tough hombre and anyone could sense that. What's more, he looked familiar. Then he introduced himself: Navy Ensign George McSwain.

I laughed and reached out to pump his hand. "Man, I thought you were dead!"

"Do I know you?" he said, laughing too.

"No, not really."

"Wait a minute, I *do* know you..." He studied my face.

I introduced myself. George's eyes lit up as recognition clicked into place. We only knew each other by sight, but in prison that was enough to imply a bond. I could tell that Larry felt left out, so I explained that George and I used to see each other around the Naval Air Training Command. He had been shot down not long after that. My buddies and I had assumed he was dead.

"Glad we were wrong, George." I grinned. "Can I ask you a question?"

"Sure. What would you like to know? "

"Who was *The Shadow*?"

Without missing a beat, George said, "Oh, that's Lamont Cranston."

"Yes, that's it! George, you just lifted a huge weight off my shoulders."

We all laughed, Larry too, as if it were the funniest thing we had ever heard.

With that, we sat down for the only real entertainment we had at The Zoo: a new story. George told us he had been an Army paratrooper in Korea before he went through the Navy flight program. He had a short career as a Navy pilot, shot down in 1966 on his first mission over North Vietnam. His A-4 Skyhawk was still over his target when the NVA shot him down and the plane exploded around him. As George put it, "My cockpit was just a piece of airplane hurled through the sky. The G-forces were so great I could barely reach my face curtain to eject." Many of us felt lucky to be alive, but in George's case it truly sounded like sheer luck.

It seemed George passed Larry's patriot test by dropping a couple of comments about his fierce hatred of communists and "gooks." After George mentioned his service in Korea and his time in college—even though he was only in college for two years, just like me—Larry was hooked. At 29, George had it all: patriotism, education, *and True Grit.* The Boy Scout also liked that George was a trove of information about engines, from motorcycles to airplanes. Whenever George spoke, Larry seemed to take copious mental notes—though sometimes I suspected he was bored and placating George. When George gave him an opening, Larry talked a mile a minute. Whatever George's heroic exploits, they always reminded Larry of something similar he'd been through. I could no longer get a word in edgewise. The way I saw it, Larry was determined to impress this intelligent, tough, charismatic guy.

At least I wasn't alone with The Boy Scout anymore.

A few days later, they moved the three of us into Alfie's old room next door and moved three new guys into our room. Our new room didn't have as big a window or a breeze, but we shrugged it off. We were used to being shuffled around.

We soon fell into The Zoo's routine. The Voice was right: rules were stricter here. However, none of the horror stories materialized. We went to quizzes, but most of the questions were social. The main purpose seemed to be to lower our morale. The interrogators said they were winning the war on the battlefield and therefore no longer needed us to provide propaganda to help them win the war in the media. The worst treatment I could complain of was being repeatedly cuffed on the back of the head for talking loudly as I walked past other rooms. It was always the same guard who hit me.

I asked George one day, "Why me? Why doesn't he ever pick on you or Larry?"

George, ever sure of himself, didn't hesitate. "Because you're the smallest."

I hadn't thought of that. "Good point."

Late that spring I came down with a stomachache that knocked me on my back. At first I thought I might have eaten some bad bananas. The camp doctor came to see me. He gave me some sulfa drugs and left. Two or three weeks later, I was still in bed, sick as a dog. I was so out of it that the time ran together the way it does in a dream.

The guards never gave me permission to rest, but George and The Boy Scout took over my duties: washing dishes, emptying our bucket, and bringing my meals. Although I felt too sick to eat, I was grateful for their kindness. For a moment, I liked Larry again. Still, I never told him I feared that whatever bug I had picked up might kill me; that I might die in this foreign shithole far from home. I thought he might find a way to use my fear against me: What kind of American didn't have the courage to face death? Worse, What kind of hero died from a stomach virus?

News trickled in from back home, though always from a Vietnamese perspective so we were never sure what information we could trust. We heard that Robert Kennedy was a front-runner for the presidency, but I didn't think much of it because I was skeptical that America would vote for an antiwar candidate.

Then one day we heard that Bobby Kennedy had been assassinated, and word got around that one of our fellow prisoners said it was too bad because Kennedy had been our best hope to end the war soon. "Mark my words," Jim Lowe said. "We're gonna be here a long time."

Back when Air Force Pilot Jim Lowe had first arrived at The Plantation, Larry was impressed because Jim had been an ace in the Korean War. But when Jim's words about the Vietnam War got back to us, Larry paced the room and said something like, "How can he say such a thing? Those aren't the words of a war hero!"

I suggested Larry cut Jim a break. "He was just expressing an opinion. Isn't freedom of speech patriotic?"

"Not in the military, not when you talk about losing. It's bad for morale."

Maybe he was right, about it being bad for morale. Jim's words did chip away more of my hope. Still, I didn't like the idea of Larry declaring what we should or shouldn't say. That night he and I got into a debate about idealism versus realism, with Larry backing idealism, me backing realism, and George acting as mediator.

At one point, Larry asked, "Are you saying you'd rather be red than dead?"

That made me feel cornered. If I said no, then I might sound as if I were backing Larry's point. If I said yes, I'd lose the argument because George was strongly anti-communist.

I refused to answer the question. The argument ended in a draw, but I could tell I had lost points with George.

During the time Bobby Kennedy went from presidential front-runner to the grave, the Battle of Khe Sanh raged on. On June 19, after more than five months, the battle came to an official end. More than seven hundred American and South Vietnamese troops lost their lives at Khe Sanh, while more than ten thousand North Vietnamese lost theirs. Yet the NVA claimed victory because the Americans abandoned their base at Khe Sanh not long after the siege was over.

Around that time, Larry was taken to quiz to talk to Dumb-Dumb. A couple of hours later, a guard brought Larry back and told me I was next.

Larry whispered to me as I left, "They want you to make a tape on Khe Sanh. Just fake it. Act like you can't read or write or talk. Screw it up."

Larry's whispers made me fear I was about to find out whether Dumb-Dumb was really the badass that Alfie had warned us about. I saw myself as a loyal American through and through, so I really did want to put up the resistance Larry suggested. Still, I knew that Dumb-Dumb would likely punish me until I did what he said. Even if I resisted, I believed that if I did what Dumb-Dumb wanted then The Boy Scout might try to destroy my reputation with George, whom I respected, as well as bad-mouth me to the whole camp. It was a lonely enough life as a POW without alienating the few people with whom I had any contact.

Just as Larry had warned, Dumb-Dumb instructed me to record a news release about the great victory the Vietnamese had won at Khe Sanh. The release was only a couple of paragraphs and, other than characterizing the battle as a Vietnamese victory, it primarily gave a body count. It was nothing more than a simple military press release. I read it, but mispronounced a few words, used odd inflection, mumbled, slurred, and generally screwed it up.

Dumb-Dumb looked exasperated. "I know you can read. You are an educated man."

I tried my best to look stupid. "Well, I only have two years of college, and I have a speech defect." Thinking fast, I stuck my tongue out and pointed at an actual slit in the surface that I had gotten when I was about four years old when I fell out of bed and bit my tongue.

He shook his head. "I heard you talk before, and you had no problem. You will read it again until you get it right. Maybe it will help you to read if you get on your knees."

I dropped to my knees and started reading again. This time I forced myself to tremble to make him believe I was scared, which I was. But I wanted him to believe I was too scared to be able to read, which I wasn't.

He clearly knew that. He slapped me hard across the face. "If you do not do this right, I will make you stay until you do, and then I will force you to read something else."

I couldn't go back to my room and tell George and Larry that I did what I was told simply because Dumb-Dumb forced me onto my knees, slapped me, and threatened me. I would sound like a coward. So I read it incorrectly a few more times, and got slapped a few more times. He sent me back to my room and told me he would talk to me again. In the meantime, he ordered me to remain kneeling in my room and ordered the equally belligerent Larry onto his knees too.

George cleared for us, standing on his bed or mine to watch out the high window for guards so Larry and I wouldn't have to stay on our knees indefinitely. Whenever George saw a guard approach, he hissed, "Down!" and we dropped to our knees. After the guard checked us and left, George watched until he was out of sight and whispered, "Clear." Meanwhile, Larry and I watched out a crack in the door.

That first day, George spent about ten hours standing on the beds watching for guards, vigilant and loyal, and we didn't get caught. But Larry didn't act satisfied with his performance.

The next morning, Dumb-Dumb ordered Larry and me to spend another day on our knees. This time Larry and George got into a debate about the best way to clear. George was moving back and forth trying to look in two directions, while Larry said he should just mind one direction while we monitored the other. They got so embroiled in this discussion that they were looking at each other and not out the window or the crack in the door. That was when I saw the shadow outside the door and dropped to my knees. There was no time to warn them, because the guard would have heard me. He flung open the door and caught Larry on his feet.

He slapped Larry twice across the face and screamed, "On your knees!"

Larry dropped to his knees, and the guard left. Larry's face was redder than his hair, and not just from the slaps. The moment the guard left, Larry berated George, "We have to have teamwork here or we're all going to get caught. If you had just done it the way I told you—"

"It's not George's fault. He was doing fine until you started arguing—" I began.

Larry cut me off with something like, "That's not the problem. Look, I'm the senior officer in the room, so if I suggest we do it a certain way, George should just do it. George has been assigned to clear this time around, so it's George's responsibility." He turned back to George. "Sorry, but the bottom line is you've gotta keep your eyes open."

I felt that he was talking to George like he was a child. Larry was, indeed, the senior officer in the room; in terms of service, he had about six months on me and two and a half years on George. But George was several years older than either of us, and had at least a year longer than either of us as a POW. In a normal situation I could understand Larry pulling rank, but this situation was as abnormal as they come.

I felt George's humiliation all the more when he responded with honest deference. "Yes, sir. Somebody has to have the final say and it's you. We'll do it your way." He agreed to keep a lookout in the direction Larry suggested, while Larry and I minded the crack in the door.

So this time, George was looking in the direction Larry ordered and Larry was still staring at George as if he were congratulating himself for his clever new system, when I saw a shadow through the crack in the door again. This time I whispered, "Get down!" It was too late.

The guard strode in, saw Larry on his feet, and cast his eyes up at the ceiling as if he didn't believe Larry could be so stupid twice in a row. "*Eo ôi!*" he said, meaning, "*Oh my goodness.*" He slapped Larry a couple of times again, ordered him back on his knees, and left.

That could just as easily have happened no matter which direction any of us were looking. None of us could clear at a one-hundred-percent success rate. We couldn't see out of our room in every direction, and nobody could keep their eyes focused for hours in one direction without shifting them. There was always a risk when getting off our knees, but it was worth it to avoid hours of suffering the pain of bone on concrete. Still, in my estimation Larry was the one who screwed up this time. He didn't say a word for hours.

Dumb-Dumb ordered us to stay on our knees for two days, with breaks for meals and sleep. In actuality, Larry and I only spent a couple of hours on our knees. But my wariness of The Boy Scout—that never ended.

Formation of A-4 Skyhawks

Hanford, CA, Fall of 1965

Christmas 1966, Cleveland, OH

Advanced flight training Kingsville, TX 1965

Robert and Pat's wedding 9 months before he was taken prisoner

Robert and Pat's wedding 9 months before he was taken prisoner

Receiving an award 1966

Aboard the *USS Hancock*; Robert is 1st row, 2nd from left

In the ready room, *USS Enterprise* 1966

NGÀY VIẾT (Dated) 30 May 1970

Mother, I received Richard's package in April and was very happy to hear from him. I could use some more soap, toothbrushes, and toothpaste. In addition hard candies would be very nice. Give my love to Patricia. Do not worry about me and tell Richard to go back to school. Love, Bob

GHI CHÚ (N.B.):

1. Phải viết rõ và chỉ được viết trên những dòng kẻ sẵn (*Write legibly and only on the lines*).
2. Gia đình gửi đến cũng phải theo đúng mẫu, khuôn khổ và quy định này (*Notes from families should also conform to this proforma*).

A short letter home

NGÀY VIẾT (Dated) 22 Sept. 1970

Dear Patricia, I would like you to buy a house
in the San Francisco area where I can go to school and
work. I would also like you to put our earnings in the
overseas savings plan and in the stock market.
Please invest very carefully. I want to meet you in
Hong Kong so please make a big list of what you
want. All my love, Robert.

GHI CHÚ (N.B.):

1. Phải viết rõ và chỉ được viết trên những dòng kẻ sẵn *(Write legibly and only on the lines).*
2. Gia đình gửi đến cũng phải theo đúng mẫu, khuôn khổ và quy định này *(Notes from families should also conform to this proforma).*

NGÀY VIẾT (Dated) 3 September 1970

Dear Patricia, I received your package

with the Princess soap in July and the

package with Dial soap in August. I plan on going

back to school to study ~~buisness~~ business and law.

Maybe you can get yourself something real nice

on the occasion of our 4th Anniversary

and your 24th birthday. All my love, Robert

GHI CHÚ (N.B.):

1. Phải viết rõ và chỉ được viết trên những dòng kẻ sẵn (*Write legibly and only on the lines*).
2. Gia đình gửi đến cũng phải theo đúng mẫu, khuôn khổ và quy định này (*Notes from families should also conform to this proforma*).

NGÀY VIẾT (Dated) 30 November 1970

My darling wife, Last month I received your April 23rd letter and picture. Boy you and Igor look so good. It made my 27th birthday brighter. I am glad to hear that you left Lemoore. I certainly hope that I will be home soon. Love,

Bob

Thanks to ex-Bay officer, POW saw his dad in time

KOSCIUSKO, Miss. — (UPI) — James Carl Bailey—who lived just long enough to see his son released from a North Vietnamese prison camp—is dead.

The father of James W. (Bill) Bailey died less than a week after seeing his son again for the first time in more than 5½ years. The father was 75.

A family spokesman said the elder Bailey died at 5:20 a.m. yesterday in a veterans hospital at Tuscaloosa, Ala. Services were today.

The younger Bailey, 30, a Navy lieutenant and pilot, returned to the United States ahead of schedule, thanks to Navy Lt. Robert E. Wideman, formerly of Bay Village, O. Wideman, a prisoner of war, gave up his place on a Feb. 18 flight out of Hanoi when he learned that Bailey's father was seriously ill with a heart condition. His mother lives in Lakewood.

Bailey arrived at Millington Naval Air Station in Memphis last Monday night and departed early Tuesday with other members of the family to visit his father. He later visited other relatives in Jackson and Kosciusko before returning to Millington late last week to undergo debriefing and medical tests.

"Seeing my son will boost my husband a great deal," Bailey's mother had said after learning her son was on his way home. "To me, Lt. Wideman is tops. This is the greatest gesture anyone could have made."

An article on Robert giving up his slot home

WESTERN UNION TELEGRAM

CLASS OF SERVICE
This is a fast message unless its deferred character is indicated by the proper symbol.

W. P. MARSHALL
CHAIRMAN OF THE BOARD

R. W. McFALL
PRESIDENT

SYMBOLS
DL = Day Letter
NL = Night Letter
LT = International Letter Telegram

The filing time shown in the date line on domestic telegrams is LOCAL TIME at point of origin. Time of receipt is LOCAL TIME at point of destination.

243P EDT MAY 8 67 AB219

A A WA194 XV GOVT PD 2 EXTRA

FAX WASHINGTON DC 8 220P EDT

ROBERT F WIDEMAN, REPORT DELIVERY, DONT PHONE
3669 DUBSDREAD CIRCLE ORLANDO FLO

I DEEPLY REGRET TO CONFIRM ON BEHALF OF THE UNITED STATES NAVY THAT YOUR SON, LTJG ROBERT EARL WIDEMAN, 689953/1315, USNR, IS MISSING IN ACTION IN NORTH VIETNAM. THIS OCCURRED ON 6 MAY 1967 WHILE ON A COMBAT MISSION. YOUR SON WAS FLYING WING POSITION TO THE FLIGHT LEADER DURING A BOMBING RUN AND WHEN THE FLIGHT LEADER WAS UNABLE TO CONTACT YOUR SON SEARCH AND RESCUE EFFORTS WERE COMMENCED IN THE AREA. AN AIRCRAFT CRASH WAS SIGHTED IN THE MIDDLE OF A VILLAGE AND A PERSONAL SURVIVAL RADIO BEEPER WAS HEARD FOR FIVE MINUTES AT THE SCENE OF THE CRASH. THE RESCUE HELICOPTER WAS UNABLE TO GET NEAR THE AREA DUE TO HEAVY ENEMY GUNFIRE. IT IS BELIEVED THAT YOUR SON'S AIRCRAFT WAS HIT DURING THE BOMBING RUN; HOWEVER THE AIRCRAFT WAS NOT OBSERVED TO BE HIT AND NO PARACHUTE WAS SIGHTED. IT IS ALSO BELIEVED THAT THE PRESENCE OF A SURVIVAL RADIO BEEPER INDICATES A POSSIBLE SUCCESSFUL EJECTION. IN THE ABSENCE OF EVIDENCE OF YOUR SON'S FATE HE WILL BE CONTINUED IN A MISSING IN ACTION STATUS PENDING RECEIPT AND REVIEW OF A FULL REPORT OF THE CIRCUMSTANCES SURROUNDING HIS DISAPPEARANCE. YOUR GREAT ANXIETY IN THIS SITUATION IS UNDERSTOOD AND YOU MAY BE CERTAIN THAT YOU WILL BE INFORMED OF ANY INFORMATION RECEIVED REGARDING YOUR SON OR ANY ACTION TAKEN RELATIVE TO HIS STATUS. I WISH TO ASSURE YOU OF EVERY POSSIBLE ASSISTANCE TOGETHER WITH THE HEARTFELT SYMPATHY OF MYSELF AND YOUR SON'S SHIPMATES AT THIS TIME OF HEARTACHE AND UNCERTAINTY. IF I CAN ASSIST YOU PLEASE WRITE OR TELEGRAPH THE CHIEF OF NAVAL PERSONNEL, DEPARTMENT OF THE NAVY, WASHINGTON, D. C. 20370. MY PERSONAL REPRESENTATIVE CAN BE REACHED BY TELEPHONE AT OXFORD 42746 DURING WORKING HOURS AND OXFORD 42763 AFTER WORKING HOURS. IN VIEW OF THE ABOVE CIRCUMSTANCES AND FOR THE PROTECTION OF YOUR SON, IN THE EVENT HE IS BEING HELD BY HOSTILE FORCES AGAINST HIS WILL, IT IS SUGGESTED THAT IN REPLYING TO INQUIRIES FROM SOURCES OUTSIDE YOUR IMMEDIATE FAMILY YOU REVEAL ONLY HIS NAME, RANK, FILE NUMBER AND DATE OF BIRTH

VICE ADMIRAL B J SEMMES JR CHIEF OF NAVAL PERSONNEL

689953/1315 6 1967 20370 42746 42763
(42).

SF1201(R2-65)

'Urgent Priority' Delays Release of Lakewood PO

"There was an urgent priority for another boy and I understand that," Suzanne Wideman said. "And I'm sure Bobby understands too."

Mrs. Wideman, of Lakewood, had been looking forward all week to the release of her prisoner-of-war son, Navy Lt. Robert E. Wideman. The Pentagon had said Wideman would be released by the North Vietnamese today.

But yesterday a telephone call from the Pentagon informed her that Robert "would not be released for reasons beyond his control," Mrs. Wideman said.

The Pentagon explained why another prisoner was replacing her son, but she was told not to repeat the reason to newsmen.

"They felt bad about it but they knew I would understand and I do. It's not easy to take but I have to take it. [illegible]," she said.

Mrs. Wideman and her husband's wife, Patricia, of Drexel Hill, Pa., have waited since May 5, 1967 when Wideman's jet bomber was shot down over North Vietnam.

Patricia Wideman, contacted in Drexel Hill, said in a distraught voice she also had been informed of the delay in her husband's release, but declined to answer any more questions.

Wideman was married six months before he was captured. "They hardly know each other," his mother said. Wideman's parents are divorced and his father lives in Florida.

Mrs. Wideman, a registered nurse on

Continued on Page 4, Col. 1

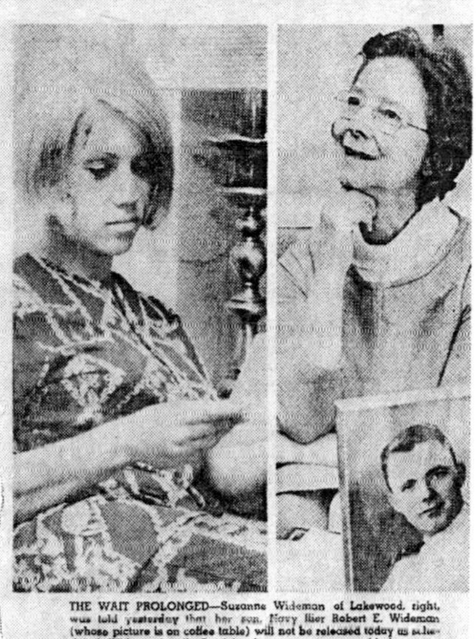

THE WAIT PROLONGED—Suzanne Wideman of Lakewood, right, was told yesterday that her son, Navy flier Robert E. Wideman (whose picture is on coffee table) will not be released today as scheduled. Wideman's wife, Patricia, shown in a two-year-old photo, continues to wait at home in Drexel Hill, Pa. The couple married six months before Wideman was captured in 1967.

Plain Dealer Photos ([illegible] and [illegible])

An article on Robert's anticipated release and return home

THE PLAIN DEALER

CLEVELAND, FRIDAY, DECEMBER 25, 1970

POW DIGS WEEDS—Man in the center of this picture, taken from a French magazine, is identified by a Lakewood woman as her son, Navy Lt. Robert E. Wideman. He is shown with two unidentified U.S. POWs in a North Vietnamese prison compound. Lt. Wideman's wife lives in Rocky River.

Long Wait All Over: POWs Write Home

By ROY W. ADAMS

Mrs. Robert E. Wideman and Mrs. Cowen G. Nix have at least three things in common:

- Each is young and pretty.
- Each is a Greater Clevelander.
- Each has a husband who is a U.S. military officer and jet pilot shot down over North Vietnam. Their husbands are prisoners of war.

The women are typical of hundreds of wives, mothers and sweethearts of U.S. ser-

year from their POW husbands, Navy Lt. Robert E. Wideman and Air Force Maj. Cowen G. Nix.

The lettrs, simple six-line messages, are all Hanoi allows. Some were written many months before they were delivered.

But they have lessened the ache in the women's hearts.

THE WOMEN, both of whom were married only a short time before their husbands left for Vietnam, work to keep their

a tree in his parachute, wearing a T-s[illegible] but smiling and saying "I'm all right."

When it is evening in Greater Clevel[illegible] it is the next day in Vietnam. And it wa[illegible] May 6, 1967, that Lt. Wideman of L[illegible] wood, now 27, took off from the deck o[illegible] aircraft carrier USS Hancock in his [illegible] hawk for a raid over Vietnam. He was [illegible] down and eventually wound up in a [illegible] compound in North Vietnam.

They had a whirlwind romance [illegible]

Robert's mother noticed him in a French magazine article about the American POW's

CLEVELAND 4-30-70

POW kin here join mail drive

By SANDY DANIELS

The movement to break Hanoi's silence about or free American boys who are prisoners of war in North Vietnam is gaining momentum.

Last week, a POW Week was observed in Columbus. On May 1, the National League of Families of American Prisoners in Southeast Asia, will rally at Constitution Hall in Washington, D.C.

H. Ross Perot, the Texas billionaire, founder of the United We Stand movement, wants to fly tons of letters to Hanoi, protesting the silence which shrouds the prisoners and camps. If they can't be freed, Perot is working with families of POWs to at least get information about and from them.

The anguish that parents and wives endure is almost immeasureable.

MRS. SUZANNE WIDEMAN, 18915 Detroit Ave., typifies the parent who is constantly hoping for news, wondering if her son is being treated humanely. She had not heard from her son in three years. She has heard of him through very indirect ways . . . people glimpsed him in films and in magazines (even a French publication).

Like Perot, Mrs. Wideman thinks the boys served their country in good faith, why should they rot in prison?

Mrs. Wideman sends packages and letters. They are never acknowledged, never returned.

"This silence is unnecessary. It is urgent to get these boys back," she says.

Robert Wideman, the POW, will be 26 years old. He was a Navy pilot, shot down, reported missing, then listed as a prisoner of war. This is all the information his mother or his wife, Pat, who lives with her parents in Rocky River, have received.

SHARING THEIR concern is Robert's father, Robert F. Wideman, who lives in Florida; his brother, Richard, who served as a helicopter pilot in Vietnam for 14 months and is now working in Louisiana.

They, with all the other families with POW sons, hope the world will get involved in getting action from the North Vietnamese officials.

Mrs. Wideman hopes Perot can get all the letters assembled to present overseas . . . the more letters, the more impact, and maybe action.

Write a concise letter to H. Ross Perot, Box 100,000, Dallas, Texas.

Maybe your letter could change the course. Maybe it would be instrumental in releasing POWs. Or, maybe it will do no more than open lines of communication, direct communication. Even that would alleviate some of the extreme anxiety of constantly worried families.

Photo of Navy pilot Robert E. Wideman of Lakewood shortly before he was shot down over North Vietnam three years ago.

Mrs. Suzanne Wideman found this photo of her son in a French magazine. It was taken inside a prison camp in North Vietnam.

ATTACK SQUADRON NINETY THREE

FPO SAN FRANCISCO 96601

Mrs. Suzanne Wideman
429 Vineland
Bay Village, Ohio

Dear Mrs. Wideman,

Just a short note to let you know how much I appreciate having Bob as a member of Attack Squadron NINETY-THREE. I am certain he has informed you of his accomplishments here aboard the USS ENTERPRISE (CVA(N)-65), but possibly he has not elaborated in detail on what I consider to be his outstanding performance as a combat pilot. During all of his many combat flight missions against the enemy here in Southeast Asia, Bob has proved himself to be a courageous and professional naval aviator. I am proud to be serving with him. In addition, his optimistic manner and cheerful attitude have also contributed much to the fighting spirit and esprit de corps enjoyed by the Blue Blazer squadron.

In closing, may I invite you to correspond with me if I can be of assistance to you or should there be any information which you'd like to have about Attack Squadron NINETY-THREE.

Sincerely,

W. G. Sizemore

W. G. SIZEMORE
Commander, U. S. Navy
Commanding Officer

DEPARTMENT OF THE NAVY
BUREAU OF NAVAL PERSONNEL
WASHINGTON, D.C. 20370

IN REPLY REFER TO
Pers G23-JDF
689953/1315

27 JUL 1967

Mr. Robert F. Wideman
3669 Dubsdread Circle
Orlando, Florida 32804

Dear Mr. Wideman:

As you are aware, your son, Lieutenant (junior grade) Robert Earl Wideman, United States Naval Reserve, was placed in a missing in action status following the loss of his aircraft on 6 May 1967 while on a combat mission over North Vietnam. A full report of the circumstances surrounding this incident now has been received and carefully reviewed. The report confirmed the information previously furnished you by the Chief of Naval Personnel, and in addition the following information was included which is forwarded to you at this time.

The weather in the target area was excellent, with no cloud cover and visibility six to ten miles in light haze. As your son's aircraft neared the target he made a radio transmission that he was rolling in on the lead aircraft's wing and he was observed visually to do so. The attack occurred at 2:08 p.m. The lead aircraft was over the water almost immediately and after several seconds he made two unsuccessful attempts to contact your son by radio. At this time an emergency survival radio "beeper" was heard on the guard frequency. The leader headed back toward the target area where two other aircraft joined him. All three aircraft spotted a fire coming from the center of a small village named XOM SE just to the east of the target. The elapsed time from the attack to the arrival of the three aircraft over the area was approximately two minutes. No parachute was observed by anyone in the flight. Search and rescue forces were alerted. After approximately 10 minutes of continued searching by the three aircraft, intense anti-aircraft fire was observed and one of the aircraft was damaged. The search and rescue aircraft arrived on the scene and they were also taken under fire. The emergency beeper was heard for a total period of about fifty minutes, however your son was not sighted during the entire search. The search and rescue efforts were terminated at 4:45 p.m. due to heavy enemy gunfire in the area which greatly jeopardized the safety of the search and rescue aircraft and prevented the rescue helicopter from reaching the scene.

On 9 March 1967, information contained in two propaganda news releases which are believed to refer to your son's capture were forwarded to you.

In view of the above information and inasmuch as the Secretary of Defense determined on 20 July 1966 that the best interest of personnel missing in action in Vietnam would be served if those believed to be prisoners of North Vietnam were so reported, the status of your son has been changed from missing in action to captured.

I join you in the hope for your son's early release from his captors.

By direction of Chief of Naval Personnel:

Sincerely yours,

H. L. Jenkins

H. L. JENKINS
Commander, USN
Director, Personal Affairs Division

Two POWs

One claims U.S. victory; one not so sure

By JOHN BREWER
Associated Press Writer

CLARK AIR BASE, Philippines (AP) — George T. Coker smoked cigars and talked quickly, almost exuberantly. A grin split his face. He said the United States had won a "fabulous victory" in Vietnam. He was ready to go back to a POW camp for American honor.

Robert W. Wideman talked quietly. He hesitated, rested his chin on his finger and thought before answering questions. He rarely smiled. He'd never go back, and he seemed dubious about what America did in Vietnam.

They were the same age—30—and both joined the Naval Reserve in 1963. Both were shot down on strike missions over North Vietnam in A6 Skyhawks from separate carriers, Coker in August 1966, Wideman in May '67. But they reacted differently in the "Hanoi Hilton" and other camps the POWs were circulated among. They were not good friends but "acquaintances," in Wideman's words. Both were released Sunday.

Coker studied German, Spanish, French and Russian with other POWs in textless seminars led by those proficient in the languages.

The Linden, N.J., resident said he also reviewed history, math and chemistry with other men and "I designed houses in my head. I put in all the lighting switches and wiring. I made eight-week menus of the meals I was going to eat when I got home."

Coker was a joiner. Wideman was not. The Drexel Hill, Pa., man stayed by himself much of the time, made three or four "good friends and no enemies" and "did a lot of soul-searching." Pause.

"What I was going to do with my life." Pause. "I was thinking along these lines before I was shot down."

What was he thinking so deeply about?

"Going back to school, getting out and accomplishing something," he said. "Most of these people—the other prisoners—were doing what they wanted to do. It was part of their service."

Long pause. "They were career military men. They made the decision to stay in and I did not."

Coker talked nonstop. He praised the camaraderie of the fliers.

"We were sent to Vietnam to do something, and we did it," he said, referring to both bombing and prison.

"I would be happy to do it again. I have no regret. We won a fabulous victory against Communism. Those who say differently simply do not understand what is going on."

Wideman also knew about the American bombing of North Vietnam. Did it do any good? "I don't know," he replied. He didn't elaborate.

When told that some POWs said they were glad about the bombing and felt that it got them out with honor, the slim, sad-eyed lieutenant paused, was going to say something, then just frowned, looking at an escort officer. The officer stopped the interview anytime it got on touchy areas such as life in the prison camps.

The Wideman Grandchildren

The Wideman Family

14

THE PIGSTY

Another choking hot summer struck just as Larry, George and I moved into a building called The Pigsty. I can't remember why we called it that, but probably because of the smell: a steaming mix of feces and rot. I had learned, though, that a human being can acclimate to just about anything. The guards put us in the end room next to the perimeter wall, where we could see our painfully close proximity to freedom. The room was smaller than the ones we had stayed in so far, but that made little difference. Prison was prison.

In September of 1968, as election season heated up back home, the daily American bombing raids in Vietnam abruptly stopped. The day the bombing stopped marked a change in the NVA's war prisoner policy, from hands-on to hands-off. Up until that point, we had lived under constant threat of punishment if we did not cooperate with the North Vietnamese. Suddenly that ceased. No more getting hogtied, slapped, or left on our knees for hours.

"It's obviously political," I said.

"Of course it's political," Larry said. "Just remember, this doesn't make them our friends." Even when Larry and I agreed it still felt like an argument.

No doubt we were both right: the Americans sought to decrease hostilities in Vietnam as a way to take voters' focus off the war during elections, while the Vietnamese didn't want news of prisoner mistreatment to slip out of the camp and jeopardize the relaxing of hostilities.

Our food improved in quality and quantity. It seemed as if the guards were eating better too. A few returned to our camp after I hadn't seen them in months, and they had gained enough weight that at first I didn't recognize them. Still, the guards only let the prisoners outside to wash dishes, shower, or go to quiz.

The national radio broadcast, called *The Voice of Vietnam*—not to be confused with The Voice of our camp—delivered news over our loudspeakers for thirty minutes in the morning and thirty minutes in the afternoon. So, although we felt cut off from America's election-season ballyhoo, and although we knew that the news we heard was tainted by propaganda, we had a vague idea that Nixon and Agnew were competing with Humphrey and Muskie in the presidential race. We also picked up a few quotes from different politicians, though Larry, George, and I all questioned whether the Vietnamese were accurately reporting what they said.

They claimed Eisenhower once said that if Vietnam had held an election according to the Geneva Accords of 1954, then 80 percent of the country would have voted for Ho Chi Minh. They also quoted from a speech on the arrogance of power by Senator J. William Fulbright, up for his fifth term. In his speech, Fulbright said, among other things, "The cause of our difficulties in southeast Asia is not a deficiency of power but an excess of the wrong kind of power which results in a feeling of impotence when it fails to achieve its desired ends. We are still acting like Boy Scouts dragging reluctant old ladies across the streets they do not want to cross."

I knew they could take quotes out of context and change their meaning. What else did Fulbright say? What else did Eisenhower say? I concluded that the quotes were probably accurate but that the North Vietnamese probably reframed them to serve a communist agenda.

A couple of times they switched the camp speakers to the *Voice of Peking* and we heard that for a few seconds. Another time, by mistake, they flipped to an American Armed Forces broadcasting station in South Vietnam. We only heard that for about five seconds, not long enough to pick up anything significant, other than the American broadcaster saying there had been a battle north of Saigon and twenty-five of the enemy had been killed with no American fatalities. We groaned when the station changed, anxious for news we could trust.

We held our own whispered elections throughout The Zoo. In the camp, Nixon won by a significant margin. We figured that our votes would reflect reality. But according to *The Voice of Vietnam* radio—and we had no reason to doubt the show's numbers because there seemed no benefit to faking them—Nixon only won by a small margin. Some Republican prisoners thought that was good news, some Democratic prisoners thought it was bad news, but most of us had no idea of its significance to us. We had been out of the loop of American news for too long. Imprisonment had skewed our perspective.

After I was first shot down, the North Vietnamese fed us reams of news reports about the massive antiwar demonstrations in the States via *The Voice of Vietnam* and the *Vietnam Courier*. Since I had heard so little about antiwar sentiment before I was shot down, and only heard about a surge in protests after my capture, I assumed the news was little more than propaganda intended to further break our morale.

Our captors made certain we heard every bit of news in which Americans or other Western nations spoke out against the war or the U.S. government. They did seem to make a distinction though, between the American government and American people. At their cruelest, I cannot recall any of my Vietnamese captors ever saying American people were bad, stupid, or unjust as a whole. They said, "Americans fought for independence, but now the American government exploits other countries" or "The war in Vietnam is bad." When they spoke about Americans as a people, they said things like, "The American people have a fine tradition of freedom." A few said to me, "The American people will rise up and convince their government to abandon this war." Was it another manipulation? Probably. At the same time, it seemed they truly admired our independence, our great economy, the American dream and all it had provided us.

Larry accused me of splitting hairs: *of course* they wanted us to hate our government so we would want to overthrow it; that was the communist way.

One night, I dreamed I was back on the USS Hancock, briefing for a 2:00 a.m. flight. The ship's lights were red to preserve our night vision. I woke up in a cold sweat, looked up at my mosquito net, and thought, "Thank God, I'm only in a communist prison camp!" I didn't fully realize until that moment that I had talked myself into believing that I *liked* flying off a carrier at night to drop bombs. In reality, I had been scared shitless every time I flew over Vietnam.

With that realization fresh in my head, I rolled onto my side, glanced in the direction of The Boy Scout's bed, and froze. An enormous rat was perched on the edge of his bed, nose-to-nose with Larry, staring at him. It had to be a foot long, not including the tail. Was it my imagination or did it look hungry? It was a robust, well-fed rat. I imagined its jaws snapping open and tearing a chunk from Larry's face.

"Larry!" I hissed.

His eyes opened and widened in horror. His body stiffened. I hadn't thought through what waking him might do. If he moved, would the rat be more encouraged to attack? Larry slowly backed away, but did not get up. The rat just sat there. I've heard people say, "They're more scared of us than we are of them," but it didn't look that way. This rat looked confident. I can't remember exactly what happened next. I may have grabbed a sandal, though I'm not sure what I would have done with it. The rat must have known a stalemate when he saw one, and waddled off into the night.

Larry sat up and put a hand on his chest. "Holy cow, that's the biggest rat I've ever seen." Even at a moment like that, The Boy Scout wouldn't swear. He nodded my way. "Thanks."

"For what? I didn't do anything."

Christmas 1968 was good to me. Pat sent a couple of items that I think she hoped would make me laugh: a quarter-pound can of Planters mixed nuts and some maroon socks that were years old. There was no note, so I'm not sure what she was trying to say: she used to call me a nut, and she used to give me a hard time about leaving my socks lying around, but I realized we didn't know each other well enough to have many inside jokes and I couldn't quite translate this one. There was also a roll of Certs breath mints, a cherry sourball, and a broken four-inch red-and-white striped candy cane, pure treasure in a prison where sugar was only a holiday event.

But the items I kept staring at were the silly bar of Avon French-milled beauty soap with a picture of Venus on it, and the black-and-white photo of my wife. I hid my face behind that picture for a while. I didn't want Larry to see the tears of joy in my eyes. Pat looked beautiful, and it seemed as if she were smiling right at me.

This picture told me she was still waiting, even though we'd now been apart for two years of a two-and-a-half year marriage. The North Vietnamese gave me permission to write my first letter home since I had been shot down.

I addressed the letter to my wife and asked her to share it with my parents. I mentioned that I hoped my brother was going to college. I was afraid to ask the question behind that: from what I recalled, Rich's military orders had been for July of 1967, just one month after I had once expected to head home. I was now a year and a half overdue. He might have already served his rotation in Vietnam, gone home, and started school. Or the Army might have delayed his rotation. Or he might be out there somewhere right now, flying a helicopter over South Vietnam's jungles, rice paddies, and bomb craters.

He might be dead.

Funny, when I tried to picture my little brother, I couldn't see him as a college student, helicopter pilot, or any sort of adult. Instead, I saw him as an eleven-year-old boy. His name was Richard Walton Wideman and I always called him Walton Ricardo. Lord, how he hated that. "Cut it out!" he would say as we walked though the woods near our house down to Casanovia Creek. If it was winter he would be wearing his ugly faded wool cap, wiping a sleeve across his runny red nose. The creek was icy. I remember the day I headed down the steep bank to the river's edge, for no particular reason except to explore.

The slope was slick with mud and ice, and I slipped in a sudden uncontrolled slide. I splashed into the water. In summer I would have swum to shore and climbed out, but I was in heavy winter clothes and the water rendered them even heavier. I tried to grab onto the bank and pull myself up, but I couldn't gain traction. The current tugged at me and I wondered if it would pull me away or if I would simply be stuck there so long I would freeze to death.

While I was struggling and making very little headway, Rich ran and found a large downed tree branch, picked it up, and returned for me. He carefully inched down to the creek's edge, leaned back against the bank for leverage, and held the tree branch out for me.

"I've got you," he said, calmly. "Grab on!"

I grabbed the branch with both hands as my younger, smaller brother pulled with all his might until the river released me and I pitched into the mud. We lay there, dazed for a minute or two, then rushed back to the house so I could change into dry clothes. I shivered the whole way back, teeth chattering. Neither of us said much, both too stunned to discuss what had just happened, but I knew my brother had just saved my life. It wasn't the last time, either.

I worried he might try to do it again, here in Vietnam, even though this time there was nothing he could do to rescue me.

On Christmas Day, our captors took us to the administration building to gawk at a small Christmas tree decorated with unfamiliar odds and ends. Of course, the tree was depressing and made me feel farther from home. But I had to give this to the NVA: they were really turning on the charm since we had stopped bombing them. This year's Christmas meal was even better than the last one, complete with a small bottle of beer.

Which brings me back to my wager with The Boy Scout. It was fifty-one days past November 4, the deadline by which Larry had bet me we would be out of the war. He was now about fifty bucks in debt to me, and he was not the kind of guy who liked to be indebted to anyone. As the holidays had approached, he'd told me that if the war wasn't over before Christmas he would give me his special Christmas meal, but he'd charge me a dollar a day for it, starting November 4. In other words, the meal would cost me the fifty dollars he owed me.

George told Larry he was crazy. It was not as if Larry would have to pay up until we were free. Until then, we were prisoners who could not afford to give up the few pleasures we received. "Jesus, we only get a few good meals a year," George said. "That's worth letting go of your fifty dollars, Larry."

The problem was, I agreed with George. Money was no object to me in prison, so giving up my fifty bucks was no big deal if that's what Larry wanted. "If you're sure, Larry, I'll be happy to eat your food for you and call it even."

"Sure I'm sure."

We shook hands. It was a bet.

But the moment I sat facing the two trays full of food sitting in front of me, and the zero food sitting in front of Larry, I felt terrible about calling in my marker. I couldn't eat that much food, even if I wanted to. With almost anybody else, I would have said, Here, take your meal back and take the fifty dollars. We'll make a clean slate. You don't owe me anything. But, knowing Larry, I felt stuck with my choice. Months in close quarters had given me some insight into his ways. If I were to admit I didn't want his food, he would win, and he would likely remind me at every opportunity that I had made a sucker bet. On the other hand, eating his meal felt like stealing food from a brother. Maybe I didn't like this particular brother much, but he was still my brother-in-arms. He knew it, too. He knew I was damned if I did, damned if I didn't. In a way, he had already gotten the best of me by putting me in this position, sitting here with my eyes sliding from my second plate to him and back again.

So I sat there and shoveled down the first meal, trying to look like I was enjoying it—but not too much—then dug right into the second meal. I got most of the way through it, too, even though my stomach was so full it hurt. I couldn't throw it away. That would be cruel. I stopped just shy of the final bites, when I felt as if I were going to throw up. I have to give Larry credit. He didn't say a word, didn't tease, complain, or beg, even though I'm sure his stomach was rumbling. We only received two meals a day, and this was one of them. I'm sure he knew I was full and wasn't getting any pleasure out of eating his meal, but he wasn't about to ask for it back, either. I did finally give him his bowl of soup. "Here, I can't finish this. Why don't you eat it?" He accepted the bowl.

That wasn't the end of it.

From then on, every rare extra goodie the Vietnamese gave us, Larry would bait me with words like, "So Bob, what would you give me for this peanut candy?" In response, I would offer to reduce his debt to me by some small dollar amount—oh yes, the clock was still ticking on our bet about when the war would end—and he would hand me his piece of homemade Vietnamese peanut brittle, which had been cooked too long and tasted gross. It seemed the easiest way to prevent his debt to me from growing out of control. The goodies were nothing special, so he wasn't giving up much, but it didn't look as if I were treating him like a charity case and giving back his money. I was already ahead, and I didn't want to break Larry's bank when he got home.

I never thought the war might go on long enough to put him hundreds of bucks in the hole.

On January 20, 1969, President Richard Nixon took office and The Zoo held its collective breath. A new president could mean new war policy. It could mean an end to my bet with Larry. It could mean an end to our lives as prisoners. Nixon immediately started talking about troop withdrawals. At first this sounded like good news, until he talked numbers. He wanted to pull twenty-five thousand troops out of the country by July. No doubt, to civilians that sounded like a lot. But George, Larry, and I knew it was a drop in the bucket.

"Twenty-five thousand troops?" I let my head fall back on my wood pallet with a thud. "Mr. President, you've got five-hundred-fifty thousand troops out here."

"It's a start," said Larry.

"Do the math. Even if they send home twenty-five thousand troops every three months or so, that'll take one, two, three, four—five years!" I didn't bother mentioning that we would remain trapped here while all those troops were leaving. It was too much to contemplate.

To me, Nixon's announcement was the first real indication we were in deep trouble. I believed that if he was willing to withdraw troops without getting back prisoners, then America was not calling the shots anymore. We were losing.

I wondered if the news reports that had filtered back to us were as out of context as I had hoped. I wondered if the quotes by all the politicians who were supposedly against the war were true. I wondered if the protests we had heard about were real after all. An idea came into focus in my mind: President Nixon was a Republican, and Republicans were reputed to be hawks, not doves. If a Republican was withdrawing troops, in whatever numbers, America must really be putting pressure on him to end the war.

Okay, fine. Maybe we had to lose this one. But if we had to go home as losers, why couldn't we do it a little faster?

Then the Paris Peace Talks started, and I felt more optimistic. Everyone did. The American bombers had not returned, and it looked like they might stay away for good. In February of 1969, the fifteen to twenty prisoners in The Pigsty circulated an invisible poll, wagering on the date we would all go home—no money involved this time.

The hope of my fellow prisoners was infectious. My guess was that we'd be freed by January 1970, only eleven months away. Ray

Vohden's guess was the most pessimistic: October 1970, twenty months away. George's estimate was as optimistic as he was: April 1, 1969, less than two months away.

Larry was unusually prudent with his guess, not as pessimistic as Ray but not as optimistic as George. Maybe losing his bet with me had taught him a lesson, or maybe he had more to lose than money or face. Maybe crushed hope was the most painful of potential losses.

All of us were wrong, about the war's end date, though we didn't know that till later. Then I would feel bad for George, who always hoped for the best and therefore seemed bound to take disappointment the worst, even though he was too tough to ever say so.

15

CRAZY

After the North Vietnamese eliminated the threat of punishing prisoners, George decided to fake insanity. He had gotten the idea from one of the prisoner-of-war training classes we were all required to take before going into a combat zone. Our instructors had told us that, during the Korean War, one American prisoner had driven his captors to distraction by pretending to have a dog on a leash everywhere he went. Amazingly, he got away with it. His captors left him alone.

"I'm doing it," George announced.

"Pretending to have a dog on a leash?" Larry or I asked.

"No. Pretending to be crazy."

Larry and I tried to talk him out of it. Relations with our captors were relatively peaceful now; why risk pissing them off again? What if they figured out what George was doing and used it as an excuse to start the rough stuff again? But George wasn't asking for our opinion.

"They're still the enemy. It's our duty to resist when we can," George said.

That argument sold Larry. I had no ground to stand on. In the end, I didn't agree to it, but I didn't oppose it either.

One spring day in 1969, our turnkey opened the door, and there stood the interrogator we called Gold Tooth Officer, or GTO for short, because half of his upper teeth were gold. Larry and I did what we always did when a Vietnamese officer came into the room: we rose to our feet and bowed. George just stood there with a catatonic stare. The only thing missing to perfect the image of insanity was a strand of drool dripping from the corner of his mouth.

GTO knitted his brows in disapproval but maintained his composure as he told the three of us, "Sit down."

Larry and I sat on our beds, but George continued to stand there and stare into space.

GTO turned to George and repeated, more firmly, "Sit down!"

George stared in GTO's direction, but looked right through him and didn't budge. GTO repeated the order to sit down two or three more times. Then he switched to asking George what his name was. No response.

GTO turned to Larry and me and pointed at George. "What is his name?"

"McSwain," I said.

He turned to George again and mispronounced his name, "Mackelson, what is your given name?"

This time George looked right at him and shouted, "Geeee-O!" at the top of his lungs. *Geo*, typically pronounced "Joe," was the Vietnamese interpretation of his name, George. Mine was *Wi*, pronounced "Why," and Larry's was *Lum*.

GTO stepped back from George's sudden burst of sound after his prolonged silence. "You have a very bad attitude," he scolded.

George didn't say anything.

"Do you know the camp regulations?"

"Nooooo." To me, George sounded more mentally handicapped than crazy.

Either way it was clearly having an effect on GTO, who stiffened and turned red. "Why don't you know the camp regulations?"

"Be-cause...they...are...stu-pid." George over-enunciated every syllable.

GTO stepped toward George until their faces were only a few inches apart, and for a moment I was sure he would slap George. But he only said, "You must memorize the camp regulations!" Then he swept out of the room.

The next day, a guard came and escorted a silent George to quiz. He was only gone for about an hour. When he returned he was covered with dust. We asked him what had happened. He said the camp commander himself had interrogated him, which wasn't all that unusual. But the commander became even more irate than Gold Tooth Officer when George refused his first question: "What's your name?" George just sat on his stool and stared through him. Like GTO, the commander repeated the question. Unlike GTO, he ordered George onto his knees. George didn't drop to his knees but just sat there. The commander left and returned with two guards, who forced George onto his knees. He went limp, and the instant they let go he slumped over. They lifted him into place again, and the commander barked at him to stay on his knees. But as soon as they let go he slumped again. Soon the irritated guards were rolling him around on the ground and kicking him.

I told George, "You see? I knew they'd start in with the punishment."

"That's the beautiful part," George said. "It was just as I suspected: they barely hurt me. They practically gave me love taps. I could tell they wanted to lay into me, but they didn't dare. The orders to keep hands off prisoners must come from high up. I just curled up in a ball and covered my ears so they couldn't damage my eardrums. No big deal."

Unable to make headway, the two guards had left and returned with two more. It was more of the same, though this time it was four on one. They forced George onto his knees, and when he slumped over they rolled, punched, slapped, and kicked him with their cheap rubber Ho Chi Minh sandals while he curled up like a roly-poly bug and waited for them to give up.

The commander left and returned with GTO. They both ordered him back onto the stool, and insisted he give his name. Again he remained silent. Again they ordered him onto his knees. Again he slumped. Again the guards pummeled him.

George never broke.

"Way to go, George!" Larry's eyes shone with hero worship.

Part of me wasn't far behind Larry. "You've got balls, George, I'll give you that." Still, another part of me was pissed. We were his roommates. What if the commander took his frustrations with George out on us?

George had worn them down. I imagined a roomful of labored breathing as GTO gave up the struggle and ordered him to go back to his room and memorize the camp regulations. George got off the stool and, instead of bowing to GTO and the commander the way we had all been instructed, he mocked the bowing routine by bowing without turning to face them as he continued out the door.

George said GTO stopped him. "He says to me, 'You must bow with respect. You must face the camp commander.'"

The three of us exchanged an apprehensive look. "I know," George said, finally showing at least some of the concern I felt. "I caused 'em to lose face. I'm probably in for it now. They're gonna call me back. What do I do now?"

"Just stop doing it," I said. "Just answer their questions."

"No. Then they win," Larry said. "Besides, then they'll want to know why he acted that way. And what's he supposed to tell them? That he was faking crazy to mess with them?"

Good question.

I recognized the look of determination on George's face. I couldn't believe it. He hadn't thought this one through, but he was committed now. He planned to keep going. God only knew how long this might go on.

The next morning, two guards flanked George to take him to quiz. Once again he was gone for about an hour, but this time he did *not* come back covered with dust. He told us that he had gone to see a fellow we called the Naval Intelligence Officer: the soft-spoken, smooth-talking, English-speaking "good cop" of the administration. He was probably younger than George, yet he took on a gentle, fatherly tone as he explained to George that he had angered the camp commander, that the commander wanted an apology, and that George must now write a letter of apology in addition to memorizing the regulations. To make matters worse, the officer then asked George to bow. George bowed all right, but it was not a normal bow. Instead of bending at a sharp thirty-degree angle, he flopped over until he was nearly bent in half, arms dragging at his sides.

George told us he didn't think he could afford to back down at this point because he'd taken his crazy-act too far. "I don't think I should write the apology."

"You can't," Larry agreed. "Once they get you to do that, who knows what else they'll try to have you do? Write a confession or antiwar statement? Make a propaganda tape?"

"That's what I thought," George said.

I disagreed. "No, no, no. They haven't made anybody do that in months. You're making more out of this than it has to be."

"Maybe, but if I keep pretending I'm nuts, then we're covered either way."

The problem was, to assure he complied with the camp commander's orders, the officer had said that one of George's roommates would accompany him to his next meeting. If he did not write the apology or memorize the regs, then George would be punished in front of his roommate. Just as I had feared: I was being roped into George's punishment though I had wanted no part of his scheme.

I figured that since I was now potentially involved, it was time to call in reinforcements to back my cause. Larry believed that George was behaving patriotically, and he was supporting him all the way. I considered his actions reckless, and I wanted to find someone wise and experienced to counterbalance what I considered Larry's hero worship. At the time, one of the guys living next door was Major James Kasler, an ace who had flown in Korea, a veteran of The Zoo who had spent the better part of the previous summer in irons, and our superior officer. I suggested we ask him for advice.

Larry rolled his eyes. "You're the only one who's unsure what to do."

I appealed to Larry's respect for order. "Major Kasler is our superior. Why wouldn't we follow his lead?"

Larry could not object to that.

I placed my drinking cup against the rear wall and used it like a mini-megaphone to explain our dilemma to Kasler. There was no crack between the walls in this room, so our only choices were either the cup or tap-code.

"They're planning on having a roommate in there with him," I said through the cup. "We think it will probably be me. Any advice for us?"

Larry interrupted, "Make sure he knows you're the one asking for advice, Bob, not us."

"Yeah, okay." I waved him off and pressed my ear to the cup for Kasler's answer.

"I think it's prudent to write the apology," Kasler said. "The stand-down is still in effect, or they would've done a lot worse to George. They're not going to pressure him for propaganda. They just want to save face."

I gave Larry an I-told-you-so look.

Then Kasler added, "My advice is to cool it, but I really can't order you guys to do anything."

It was Larry's turn to look smug.

We got off the wall and huddled again. Larry and I knew neither of us would budge, so we left it to George.

I didn't anticipate his decision, which none of us had recommended: "Next time they come to take me to quiz, I'm just not going with them at all."

This dumbfounded both Larry and me.

Sure enough, two guards came back that same afternoon and told George to get ready for quiz. George was standing when the guards entered. Getting ready for quiz meant changing out of the t-shirt and shorts we typically wore in our room and into our pajama-like long-sleeved prison smocks and pants. Instead, George retreated to the far corner of his bed and sat there.

"Go to quiz!" one guard insisted. Then both of them went outside to wait.

A few minutes later, they returned. George hadn't moved, except to press himself more firmly against the corner of his board and the wall. One of the guards tried to nudge him off the bed. George scuttled backward like a crab. They grabbed him, dragged him to his feet, and prodded him. Each time, he pulled away, slumped, or retreated back to the bed. The guards' astonished expressions told us that, out of all the crap they had dealt with, they'd never seen a prisoner simply refuse to go to quiz at all.

It occurred to me at that moment that I had rarely seen a guard punish an American in front of another American—maybe a slap at most, but nothing heavy-handed. If that was a camp policy, then these guys were in a pickle, because George wasn't going anywhere and Larry and I were witnesses.

After a few more tries, the guards exchanged baffled looks, shrugged, and departed.

That evening, the Naval Intelligence Officer paid us a visit. Once again, George just sat on his bed in what Larry and I dubbed the "vegetable position," while Larry and I bowed. The officer sat on Larry's bed, crossed one leg over the other, lit a cigarette, looked up at Larry and me, and gave the wooden board a pat, like a dad patting a mattress to invite two sons to a heart-to-heart talk.

"Please, sit down," he said to us.

Larry and I sat on either side of him, and for the next thirty minutes we listened as he explained to us the trouble with George, who continued to stare unblinking into space as if unaware of our presence. George had made the camp commander angry, the officer said. He described everything that George had already told us about his quiz, minus the beatings of course. He even got up and demonstrated the way George had bowed.

"What do you think when he bows like this?" He slumped over, letting his arms flop like noodles and his fingers drag on the floor. Then he stood back up and gave us an earnest *Father Knows Best* sort of look.

God, it was funny. I didn't dare look at Larry for fear I would burst into laughter. I held my lips straight, my eyes wide.

"What do you think?" he repeated.

We nodded slowly, as if to say, Hmmm, we see what you mean.

"Your roommate has a very bad attitude and the camp commander is very mad, but if the camp commander gets an apology, then everything will be okay." The officer bent over his knees as if getting ready to push up from the bed and leave. Then, without warning, he rose with his rubber sandal in hand, swung around, and smacked George full force across the face.

Larry and I started at the surprise, but George did not blink an eye.

The officer studied him for a moment. Then he put his sandal back on and left.

George dropped the potted-plant act and we huddled up, with Larry clapping him on the back for standing his ground. My arguments didn't stand a chance.

The next day, they called me out for a quiz, and I found myself standing in front of the camp commander. His face warned me not to screw with him, but his voice remained calm.

"Your roommate, George, is he well?"

So, I thought, he's going to play along with George's own game to try to trip me up. I chose my words carefully. "He's okay...but he's different."

"What do you mean?"

"He's hard to get along with. He doesn't talk much to Larry or me, and when we do talk he's...different."

When he couldn't get me to elaborate, he sent me back to my room.

I told George what had transpired. "Beautiful!" he said, clapping his hands silently as if this was going just as he planned.

"Whaddaya know, they really believe he's crazy," Larry said.

I nodded. George's ploy seemed to have worked. Something felt off about the commander's reaction, but this time I really wanted to be wrong.

It might have bolstered George's case that there was another guy in the camp who actually was suffering some sort of breakdown. We heard reports through the walls that he wouldn't eat and he was melting away to nothing, weighed maybe 110 pounds. We heard that he wouldn't bow either and that the guards hit him across the face with a Ho Chi Minh sandal and he wouldn't even blink. We heard that it wasn't just the Vietnamese he had stopped talking to, that he had clammed up in front of his eight roommates too. He had received a photo of his wife and kids the previous Christmas, and when he descended into silence his roommates propped that picture on his bed to lift his spirits, but he took one look at it and knocked it to the floor. I don't know how much of what we heard through the walls was accurate, but I later heard that the NVA took him to a hospital, where he died without ever seeing his wife and kids again.

George was too stubborn to truly fall apart like that. We only hoped the Vietnamese didn't know that.

The day after the commander questioned me, he called Larry in for a quiz that went more or less the same as mine. In the coming days, the commander began alternating quiz days between Larry and me. He asked us for daily reports on George's behavior. Soon he began asking us to write our answers: *Today he got up and brushed his teeth... He would not wash his clothes today. We washed his clothes... He would not go out and bathe... He has not shaved now for two weeks.* This time the three of us worked together on a plan to use these reports to our advantage. George knew he was out on a limb with his bogus mental illness. The problem was, he couldn't pretend to get better all of a sudden. Then they would know the truth. So we decided to paint a picture in which George would get worse for a few weeks, then stabilize, then get better, and finally apologize to the camp commander.

The guards stopped punishing George. We talked ourselves into the hope that our ploy was working and the hope that perhaps this was another sign the war would end soon. Maybe they feared sending home a crazy POW because then the world could say, "Aha! You treated prisoners so badly they went insane."

The interrogator asked Larry and me to give increasingly personal details about George's life. Some of the stories we told him were true, some weren't. It was hard to tell if we were fooling him or not. His only reaction was, "He is your comrade. You must help him." We could not interpret what that meant with any certainty: either they believed George was going loony and really were begging us to help turn him around, or they knew it was a hoax and were giving us a chance to convince him to come around on his own so both sides could save face. Either way, we knew there was a limit to how long they would give us. Either way, we all had a stake in this game now: whatever happened to George would impact all of us.

Seven weeks passed. It was time to change our story: "We think George is getting better... He ate his breakfast this morning... This morning he said hello to us for the first time in two weeks." We just needed maybe a couple more weeks to declare George cured and it would all be over.

One day in May, I came back from quiz just as Larry left for quiz. George was lying on his bed, still playing possum. A faint sound broke the silence: Kasler tapping a staccato message on the wall, the usual signal that our neighbors wanted to communicate: *Shave-and-a-haircut, two-bits. Tap-tap-ta-tap-tap, tap-tap.* We knew the guards might bring Larry back any moment, so it seemed safer to tap than to talk through a cup against the wall. I tapped what had happened during my quiz. Kasler was tapping a reply when I heard someone open our door-flap. I looked up and saw Gold Tooth Officer staring at me where I stood next to the wall, listening. His face tensed and then vanished. I knew he must be heading for Kasler's door on the other side of the building to catch him in the act. George heard GTO running around the building, so he reached up from his place on the bed and thumped the wall twice, to signal Kasler that danger was coming.

What I saw and George didn't—because his back was to the door—was that a second guard had just walked up to the open flap in our door, right on the tail of GTO's departure. That second guard saw George pound on the wall, and his look said it all: If George was catatonic, how had he known to warn our neighbors? His look said, Gotcha!

I could almost taste my heart in my mouth. The guard opened the door just as GTO returned. He and the guard spoke in rapid-fire Vietnamese for a moment. GTO's eyes lit up and then narrowed, not at George but at me. "Ooh, you have made a very big mistake. Put on your long clothes. You're going to quiz."

My bladder felt like it was rapidly shrinking. We had put our captors through this charade for months. How far might they go to reassert control?

At quiz, GTO himself interrogated me. "What's the story?"

"Gee, I don't know." Playing stupid seemed like my only option. Maybe he would believe that George had pulled one over on me too, or maybe he would believe that George really had been getting better and this was just another phase in his miraculous recovery.

"Who is the man next door?"

"Kasler."

"What were you talking about?"

"We were talking about health and children and family and school." This was the cover all the prisoners had agreed to give whenever we got caught communicating.

"No," he said. "The other man confessed: he ordered McSwain to be crazy."

"What? No." Of all the lies we told, that wasn't one of them. If anything, Kasler had tried to talk George out of the crazy-act. But once an interrogator got an idea in his head, there was no way to change his mind. I already knew, the more I denied the accusation, the more it would look as if I were covering something up. But I couldn't afford to confirm what he said either. I definitely couldn't admit I knew that George wasn't crazy.

"Kasler is superior officer," GTO insisted. "He ordered McSwain to be crazy."

"No. I'm telling the truth. Kasler and I only made small talk, and we talked about how he broke his leg when he crashed. He did not order anything."

GTO ignored me. "Time for dinner."

He didn't send me back to my room to eat. Instead, he left me alone in his office, and a guard brought my dinner there. GTO returned at sunset with four extra guards.

"You will move to new room tonight." His voice was unusually pleasant. I had never heard him sounding so cheerful, and it scared the piss out of me.

GTO and a phalanx of guards led me into the evening sun, past the parade-ground-sized courtyard with its algae-slimed swimming pool, to one of two small rooms on either end of the Auditorium, a large building that faced the courtyard with its paved roundabout. The camp commander stood off to the side, long arms folded, features downturned as usual, which made him look to the prisoners like a baboon. Some people called him that, The Baboon.

The room was empty of anything save a set of leg irons, a plank bed, and a shit bucket. The guards ordered me to sit on the floor while one of them clapped my legs in the irons. They hauled me onto my knees, a procedure that was even more awkward than usual, what with the shackles around both ankles and the two-foot iron bar between them. Then GTO stepped in and gave me his parting shot, "You must tell the truth. It could be very, very bad for you if you do not tell the truth."

In theory, I was supposed to remain kneeling at all times except bedtime, until somebody excused me. But as soon as GTO's footsteps receded, I rose to my feet to give my knees a break and have a look around. Oddly, my new room seemed slightly better than the one I shared with George and Larry, narrower maybe but longer, with a slight breeze from the high barred window and louvered door. What's more, the louvers allowed me to see into the courtyard to the east. Not only was it a nice change of view, but it also made it easier to see the guards coming so I knew when to drop to my knees. If I hadn't been so terrified about what the next day might bring, I might almost have enjoyed it. I had forgotten how peaceful it could be to have time and space to myself. That is, until night fell and the mosquitoes swarmed.

The room had no mosquito net. Mine was back in the room I shared with George and Larry. How could I have taken it for granted, the thin cocoon of fine olive-drab mesh that was in reality a critical piece of military armor? I soon found myself at the bloody, itchy center of a gray cloud, pacing and swatting at myself like twice the madman George had pretended to be. The moment I got rid of five of them, another ten would land. Maybe I *would* go mad after all. I tried to lie on the ground near the front window, where I thought the breeze might keep the mosquitoes at bay. It didn't help much.

I didn't sleep all night. Just waited to see how bad GTO's "very, very bad" might be. Whether I told the truth or confessed to the lie he insisted on, neither was likely to protect me.

I was wide-awake when dawn broke, and I jumped up at an unfamiliar commotion outside. I peered into the courtyard through the louvers in my door, which split the scene into dozens of horizontal lines. That and a lack of sleep made what I saw next seem more like a low-budget TV show than reality. I saw a couple of Vietnamese guards march an American prisoner, manacled and blindfolded, from the direction of the part of camp we called the Annex. They passed the administration building, known as the Head Shed, and went out of sight. Then the guards crossed back and marched out another manacled and blindfolded prisoner, and another. One by one. Six or seven in all.

That was maybe the strangest thing I had seen since the bombing had stopped some six months earlier. Up until that moment, it was both the most punishing and the most relieving thing about prison life: nothing exciting or unusual ever happened.

16

NO ESCAPE

I spent most of the morning standing in my leg irons and staring through the louvered door as I scratched the red mosquito bites scattered over my arms, legs, neck, chest, and face—almost every inch of me. I looked as though I had a bad case of chicken pox. The mosquitoes had retreated with the rising sun, but I knew they'd be back. For now, I was more interested in the manacled, blindfolded prisoners the guards had just escorted past the Head Shed.

It wasn't long before another mystery arrived. The front gate, which I could see from my new cell, opened. A small sedan drove in, circled the roundabout, and came to a stop in front of the Head Shed. A rear passenger door opened, and a Vietnamese man stepped out, dressed in civilian clothes: dress slacks, white shirt, rolled-up sleeves, no coat. Another civilian and a military officer in an impressive uniform stepped out of the car too. All three had the air of men used to giving orders, and the man in uniform had the look of someone the world made way for—orders or no.

The trio walked straight toward my room. For a moment, I wondered if I was in deeper trouble than I had thought for my part in George's rebellion. They came to a stop twenty feet from my door, squatted on the pale gray asphalt of the roundabout where it passed in front of my building, and conferred in hushed tones. They probably had no idea anyone was nearby or they would have talked elsewhere. Not that I could understand what they were saying. Was it about the prisoners I had just seen?

As if in answer to my unspoken question, the officer pointed to a corner of the Head Shed, where the guards had marched the bound prisoners, then pointed toward the Annex. He drew something in the dirt. The three men stood up, and the head honcho's eyes and shoulders rose and fell in a show of disgust. Then they went their separate ways, walking to different parts of the camp.

Nothing much happened for the next hour or two. Then the gates opened again. This time a jeep drove into the camp. I had a hard time following its progress because it seemed to be moving erratically, and at one point a group of trees blocked my line of sight. Next, I heard a crash. A bunch of guards ran in the direction of the trees. I had the impression that beyond those trees the jeep had hit a building. The guards were all talking and yelling at once. One stepped away from the others and called out for help. Someone must have been hurt. A couple of medics ran across the courtyard and through the trees. They reemerged carrying someone away from the crash site to the Carriage House near the gate.

Shortly after that, I saw a guard leading another blindfolded prisoner past the Head Shed. This prisoner was covered in mud. He was wearing something that looked to me like a straightjacket, though maybe his hands and feet were just tied behind him with a complicated system of ropes. About twenty minutes later another guard ran another muddy prisoner across. This time I could make out reddish-brown hair through the dense coating of mud but could not distinguish his features. He limped and gasped as if in pain, though maybe that was just because he too was in some kind of straightjacket. I was convinced that these two prisoners must be new shoot-downs, covered in mud from the rice paddy, river, or swamp where their planes had crashed.

Not long after that a prisoner emerged from the far side of the Auditorium, the building I was in. Although he too was blindfolded, he was not tied up like the others. He was carrying the rolled-up mat, cup, and other standard items doled out to prisoners. His height, profile, and distinctive receding hairline—which seemed to retreat in the opposite direction of most guys, from back to front, instead of front to back—all looked familiar. He looked like my first roommate, Mike McCuistion. Before I could be sure, before I could piece together a clear image of him between the slats of my door, they shoved him into the room on the opposite side of the Auditorium from me. There would be no way to ask him what was going on.

Was Mike being punished for an infraction? Were the guards moving him to make way for new prisoners? Was he in trouble because he was my old roommate and the guards believed we were both part of a complicated conspiracy that had culminated with George McSwain acting crazy?

Maybe it wasn't Mike.

That evening, all the North Vietnamese who worked at the camp lined up in rows in the courtyard: interrogators, guards, turnkeys, medics, grunts. I was surprised at how few there were, maybe fifty all told. The air filled with voices that hummed with tension. The head military honcho, whom I mentally nicknamed Tojo, swept out of the Head Shed to the front of the gathering, and everyone fell silent. He clasped his hands behind his back and spent the next half hour delivering what could only have been a stern lecture.

During Tojo's lecture, a guard the prisoners called Magoo walked past my room and then came back the other way carrying a bunch of leg irons. The leg irons I was wearing were the only ones I had seen for at least six months, ever since the camp standdown on severe punishment. That's when I knew something serious was up.

After Tojo dismissed everyone, he led the camp commander and the Vietnamese Air Force liaison for prisoners across the parade ground to the algae-slimed swimming pool. There, atop the shrubs next to the pool, sat a pile of prisoners' gear: clothing, mats, mosquito nets, cups. They lifted up the items one by one, inspecting them, making an occasional comment, shaking their heads. One of them lifted something small between two fingers and wagged it in front of the other two men. It looked like a bottle of iodine. The others smirked as if at something ridiculously stupid.

It was the most exciting day I had witnessed in a monotony of months. Like a bedridden single guy stuck in his apartment watching bad TV via an antenna that receives only one channel, I could not turn away.

Kitty-corner across the courtyard sat a building that didn't usually hold prisoners as far as I knew, but today I could see through the barred window of one of the rooms. Two blindfolded prisoners were sitting on plank beds with their hands tied behind their backs. One of them had his legs crossed and clasped in irons, which I knew made it impossible to find a comfortable position. Still, they appeared calm. Meanwhile, at the nearby Head Shed, a prisoner's water jug and rolled-up bamboo mat sat on the porch, indicating that a prisoner was likely being questioned somewhere in that building.

Soon an ambulance pulled into the camp and backed up alongside the Head Shed. I couldn't see the rear of the vehicle, but it sounded as if men were loading it. Then the ambulance rolled away in silence—no lights, no sirens. I saw that the water jug and bamboo mat had vanished from the porch. I speculated that a prisoner had been taken to the hospital. I hoped the only reason for no lights or sirens was because he was not badly injured, or that they didn't want to draw attention, not that he was dead, whoever he was. Any one prisoner's fate always seemed inextricably bound up with all our fates. It was hard to see each other's lives as separate.

***

That night, the mosquitoes returned in force. Who needed to fake madness? The dark cloud of miniature torturers drained my will to remain civilized. In desperation I took the lid off the shit bucket and rubbed my own urine and feces on every inch of exposed skin—face, hands, feet, everywhere—hoping it would act as a natural insect repellant. I stank to high heaven and my already sweaty skin felt clammy, but my do-it-yourself insect repellent worked...for about twenty minutes. Once the excrement dried, making my skin feel stretched like tanning leather, the mosquitoes returned. Their proboscises left a trail of welts between the cracked and hardened smears of my own waste. I slapped at myself, adding splatters of blood to my coating of bodily execrations. I lay on the floor as close to the window as I could get, hoping for relief from the faint breeze. The mosquitoes were undeterred.

The guard looked in and saw me lying near the window. "You must go to the bed in back of the room."

I rose reluctantly, shuffled to the back, and dropped onto the plank that served as my bed. I don't think the number of mosquitoes changed. They simply followed the food supply wherever it went. I didn't sleep all night. At dawn, I pushed up onto my still-shackled feet to see if the show I had witnessed the previous day would continue.

The guards seemed on a new mission to spread order, moving en masse throughout the many buildings of the camp, which was pretty unusual. About an hour later, the guards filtered back to the Head Shed. One of them, the Naval Intelligence Officer, was holding a paperback book, one that I recognized. It was a book published in East Germany, titled *The Hundredth Century*. Like so many things since I had met The Boy Scout, it had been the source of an argument between us.

Larry had stolen the book from the North Vietnamese about three weeks earlier. I had berated him for taking it upon himself to do something that could get all three of us in trouble—him, George, and me—without asking the other two of us whether we were willing to share the risk. Larry had given me his usual rigmarole about it being our duty to disobey and resist the enemy.

"You lie in your bed and don't do anything," Larry said. "At least George and I are doing something to resist the Vietnamese. You should be grateful. You're always saying you're bored, and now we have something to read that we didn't have before."

"It's kind of ironic don't you think?" I said. "Here you are, an American in a communist camp and what do you do? You go out and steal communist literature to read. All you're doing is stealing the propaganda they want us to buy into."

The book was written in English, so of course we all read it. Stir-crazy as we were, if it had been a book about air conditioning repair we still would have read it. As it was, air conditioning repair might have been better. *The Hundredth Century* contained about two-dozen of the most depressing short stories I've ever read. The stories were meant to be object lessons about bad things that can happen in life. None of them featured the good side of human nature. None of them had happy endings.

As I watched the guard take Larry's book to the Head Shed, I knew we were in deep trouble now, much like the people in the book. A moment later, another guard followed, swinging a small English-Vietnamese dictionary at his side, another item The Boy

Scout had stolen from an interrogator and hidden behind the folded-open shutters in one of our windows.

For the next hour, I waited for someone to slam into my room and march me off for questioning. But I had no energy to sustain panic. After my wakeful night with the mosquitoes, I was dozing on my feet. Nobody came by my room to order me back onto my knees so I took the chance to curl awkwardly on the floor near the window and catch up on a few Z's.

I woke to the sound of the door opening, scrambled to my feet, and bowed to GTO. I waited for him to say something about the books.

Instead, without preamble, he asked, "Why did Kasler order McSwain to be crazy?"

"I don't know anything about that."

"No. You lie. I know."

"I'm telling you the truth. I never heard Kasler order George to do anything. George doesn't talk to us either. He's very hard to live with."

"No. You must confess correctly." He swept out of the room.

Moments later an interrogator we called OJT came in. OJT was short for On-the-Job Training. We called him that because he was young, new, and nervous, and always seemed to be shadowing a more experienced interrogator.

"Are you ready to confess?" OJT squeaked.

"No."

"Okay, now you will be punished." He forced me to my knees, which smarted because my leg irons jolted me to the ground so that my knees hit the concrete with a crack. He stalked out.

Two seconds later GTO returned to say, "See, now you are being punished."

It was as if they were playing good cop/bad cop but nobody had explained "good cop" to them. GTO told me the punishment would not stop until I confessed. Then he strode out again.

At first I thought the punishment wouldn't be too bad. Through the louvered door, I could see the guards coming and going, so I was able to rise from my knees occasionally and clear for myself. Then, in the evening, I got caught off my knees. Four guards charged in and rolled me around on the floor, kicking me with the sides of their feet. This caused no pain. I told myself it was merely annoying. I didn't think about it being humiliating; I was beyond all that. One of them pulled off my rubber sandal and hit me square in the nose with all the force of a fist-punch, bringing water to my eyes. I raised my hands to protect my eyes, worried the guard might miss my nose and do one of my eyes serious injury. How would I ever become an airline pilot when I got out of here if my sight were damaged? Even at that moment, I was trying to think past being a prisoner, as if this were a dream and soon I would resume my waking life.

The beating stopped as suddenly as it started. They hauled me into a kneeling position and ordered me to stay put. This time they left the door open and posted a guard to ensure I stayed on my knees. But the guard was off to the side of the doorway and could not see my every movement. I leaned over and pushed the louvers to a more optimal angle, so I could see him whenever he made his move to walk over and peek in. I was able to continue clearing for myself, only dropping to my knees when the guard approached the door.

I wasn't caught again. Or at least, if anyone saw me off my knees they didn't think it was worth the effort to keep giving me half-hearted beatings.

The puzzling parade of shackled prisoners and angry guards never returned. Nobody told me what the strange day of commotion the day before had been about, and I didn't ask.

GTO returned day after day to interrogate me about George and Kasler. Sometimes he expanded the scope of his questioning to include our camp-wide communication organization: "What is the chain of command?" "Which rooms communicate with each other?" "How do messages go from building to building?" "Who are the key people to carry messages?"

"I don't know. I'm just in one part of one building, and I only communicate with the people next door, and only if they ask. Most of the time they don't want to communicate, because they're scared of being caught."

"What do you communicate about?"

"Family, health, what they plan to do when they're released."

"No. Not true. You must confess correctly."

He was right. It wasn't true, but what could I do? If I denied we had a communication system when they clearly knew we did, they would assume I was also lying about George and Kasler. If I admitted we had a communication system, it would confirm their suspicion that George and Kasler were part of a bigger conspiracy of disobedience. So I rode the middle ground.

I waited for GTO to ask me about the stolen books, but he never did. I suppose he already had Larry dead-to-rights on that one and felt no need to question me. I had no idea what was happening to Larry or George. I knew it was pointless to ask.

Although I was in solitary, I wasn't entirely cut off from the prisoner communication system that GTO was dying to decode. Sometimes prisoners walked by and whispered, "Hi." Sometimes they asked my name, how I was doing, or how the "gooks" were treating me.

One evening after dinner, I used my aluminum spoon to carve a message into the bottom of my metal plate. I carved the following words into the white enamel: *Tell McSwain & Kasler I'm O.K. In Irons. Admitted "trying" to talk with Kasler, but said he would not answer. – Wideman.* I knew that any prisoners assigned dishwashing duty always checked the bottoms of plates for messages. They would pass my message on if they could.

After my next meal, my guard told me to scrape the meager leftovers from my plate into my shit can and then set the plate outside the door. Anxiety pierced my stomach. Did they find my note from the previous plate? Luckily I hadn't written a note this time. I didn't know if the first message had gotten through, but I did not take the chance of doing it again.

For the next few days, my only communication was with my interrogator, to whom I continued to reveal nothing. Despite the nerve-wracking interrogations, I was relieved to get away from the constant tension with Larry. I had almost forgotten what true privacy felt like. It was the closest I could get to freedom in this place. Although I would have liked someone else to talk to now and then, I liked having a room to myself, or almost to myself—if only I could get rid of the mosquitoes.

Years later when I heard Vice Admiral James Stockdale say that solitary confinement was the worst punishment of all, I laughed. What I would have given to spend more time in solitary! To me, it meant peace and privacy, not being force-fed the opinions of others, not being held accountable for the actions of others. While I was in solitary I remained in leg irons 24/7, but at least my legs weren't crossed. The mosquitoes kept me awake every night, but every sun-up they departed and I could squeeze in a nap. The guards ordered me back onto my knees every day, but I only spent a total of about twenty minutes a day on my knees, dropping to them only when I saw the guards coming.

Every day, GTO came by to insist I admit that Major Kasler ordered George to act crazy. Every day, I insisted that George acted crazy around us too and that Kasler had ordered nothing.

I figured that, so long as I could snag a few hours of sleep each morning, I could hold out for months without fear the enemy could coerce a confession from me. Although I had objected to George roping me into his act of defiance, I still admired the guts it took for him to carry things this far. I had to admit he had the grit of a real American hero, and I wasn't about to betray him if I could help it.

17

CAPTURING THE TIGER

I stayed in solitary confinement for two weeks. One day GTO returned with several guards to let me know that, whether or not I'd had enough, he sure had.

He stood over me, on my knees once again, still hobbled by leg irons, and commanded: "You must confess that Kasler ordered McSwain to be crazy, and you must confess that you violated the camp regulations, and that you communicated, and that you made noise in your room, and that you tapped on the wall."

I looked up at him and saw in his eyes that he would do whatever it took to achieve that result. All right then, if he was going to demand the truth, so was I. "I will confess to all of the things I did, but I cannot confess that Kasler ordered McSwain to be crazy because it's just not true."

We measured each other for a moment. Then he came to a decision. "You will confess that you communicated with Kasler."

I nodded. He nodded at the guards. One lifted me to my feet. The other handed me a pencil and a sheet of paper. I sat on my plank bed, set the paper on it, and scribbled a confession while GTO stood over me. In my written confession, I admitted to most of what he said but kept it as vague as possible. When I got to the bit about Kasler, I wrote, "I communicated with a man next door and we talked about family, health, and what we plan to do when we go home."

I signed it. GTO yanked it from my hands and walked out, leaving me sitting on the bed and the guards standing around me, all of us waiting. He returned about half an hour later.

"Your confession is not good enough." He slapped a paper next to me on the bed. "Here, you must sign this."

It was the confession I had written, with an additional two or three sentences written on it in broken English. The new sentences declared that I confessed to violating camp regulations by communicating with a man next door, that I was sorry, that I wouldn't do it again, and that I would obey camp regulations in the future. There was no mention of what Kasler and I had talked about. I signed it. GTO snatched it from me and walked out.

The guards removed my leg irons and walked me back to the room I shared with George and Larry. I didn't know how to feel about this. It seemed my punishment was over, but there was no way to be sure.

The first thing I noticed was that our room was buttoned up. Our shutters, behind which Larry had hidden the Vietnamese-English dictionary, were now closed. So there was no longer any way to tuck something away behind them. The room felt sauna hot, making me wish even more that they had just left me in the other room with its wisp of a breeze.

The moment the guards left, I turned to George and Larry, expecting a lot of questions, but hoping for a few answers of my own.

"Have you heard the news?" George said.

He had my attention. News was a scarce commodity. "What?"

"Did you hear about the escape?"

That single word, "escape," unlocked the mystery I had seen unfold in the courtyard the day after the guards threw me into solitary. Larry and George weren't sure of the details, but two prisoners had attempted an escape: Ed Atterbury and John Dramesi. My roommates weren't sure what had happened to Ed and John, but the more critical information for us was that their six or seven roommates had been trussed up in ropes until they confessed.

"They told the Vietnamese that our room was the 'ready escape room,'" Larry said.

"Our room?" My stomach lurched.

We had known nothing about any escape, but either somebody had fingered us in a panic, or the Vietnamese had suggested the idea because they hoped it might explain George's sudden madness—maybe they thought we had all cooked up George's little act as a diversion. Either way, it made me wonder why the Vietnamese didn't use that as an excuse to come down harder on me during my time in isolation. I felt lucky, grateful even.

"Although they didn't get away, hey, at least they gave the guards something to think about," was Larry's assessment.

I thought about the resulting shakedown that had unearthed Larry's stolen books. If the escape had given the enemy something to think about, it had also put all of us on the line. But I didn't say that, just nodded as if taking in the weight of Larry's words. Then I changed the subject. "So, did you guys get my message?"

"Something got back about you in irons, and then we lost communication with you," George said. "We were concerned about what was happening to you. We didn't know." His face told me that they both had truly worried whether I was being tortured. Whatever else we felt about one another, none of us wished punishment on our fellow prisoners.

One thing all three of us agreed on, now more than ever: George could not afford to suddenly return to sanity.

It seemed that the safest course of action at this point was for George to act even more mentally unstable. If he were to snap out of it, not only did we risk looking like liars, we risked implicating Kasler as the one who really had ordered George's bout of insanity. So George kicked it up a notch, behaving increasingly loony and disconnected from reality. He completely stopped shaving, washing his clothes, or bathing himself. He hadn't shaved for weeks. Larry and I washed his clothes. Larry and I took George by the hands like a child, led him to the large concrete shower-room in our building, and pretended to wash him, all under the watchful eyes of the guards.

I had been back in our room for a week or two when I was startled by the sound of someone tapping our old signal on the neighboring wall. In the wake of my punishment and the capture of the escapees, we had all been cooling it on communication. So the *Shave and a Haircut* knock, *thud-thud-tap-thud-thud*, probably would have made me jump in any case. But something was screwy about the rhythm and the tapping was unusually loud, which no prisoner ever risked. My roommates and I sat up and exchanged nervous looks. We turned to stare at the wall, but nobody moved to reply.

A second later, the flap in our door opened and there stood this fellow we called The Rat, an English-speaking Vietnamese with a nose that reminded us of a rat. "Take the message from your comrade."

George's face went blank, as it always did these days when there was a guard in the vicinity. The three of us sat stock-still, barely breathing.

The Rat opened the door, walked in, and handed me a piece of paper and a pencil. "You must take the message from your comrade."

I studied his face for a clue to what was going on, but his expression revealed nothing. I accepted the writing utensils. "Okay."

Someone tapped a message, and I took it down letter by letter until I was looking at a sentence something like this: *McSwain, show a good attitude and go to quiz.*

The tapping stopped. I stopped writing but did not look up.

"What does it say?" The Rat asked.

I stared at the paper, afraid to say too much, afraid to say too little. Just how much did they know about our communication system? "I'm not sure. It was hard to understand. Something like, 'McSwain, go to quiz.'"

He gave me a firm nod, and turned to look at McSwain, who continued to stare into space. "Yes, McSwain, go to quiz. That is an order from your comrade, Major Kasler." He walked out without another word.

A while later, the *Shave and a Haircut* tap came again, softer this time, more like the signal we knew. Kasler tapped out a new message. He explained that the Vietnamese had taken him out and given him six lashes to the butt with a fan belt, a punishment we had heard of before, though not in the months since the stand-down on strict punishments. Kasler said the Vietnamese had instructed him to send us the message ordering McSwain to go to quiz.

The next day, the same thing happened again: knuckles rapping a big hairy *Shave and a Haircut*, the flap opening, The Rat's nose appearing, the door opening, The Rat handing me paper and pencil, saying "Take the message from your comrade," Kasler tapping out a message.

This time the message read, "McSwain, put clothes on. Go to quiz."

Clearly, they really were convinced Kasler was the one who ordered George to act crazy, and now they thought they could use the same person and same communication system to order him to stop. If anything, this had the opposite effect. As soon as The Rat was gone, Kasler tapped out a new message: "Whatever you do, don't go to quiz now because it really will look like you're following my orders. Then they can say I also ordered you to be insane."

"He's right," Larry said.

I gave Larry an irritated stare, thinking, Yes, we all know that, Larry.

Not long after I took down the message ordering McSwain to quiz, two guards opened our door. That wasn't alarming in and of itself because it was our room's turn to hit the showers. Larry and I kept up the usual act, leading George to the showers, soaping him, and then leaving him standing dumb and limp-armed in a stream of water while we bathed ourselves. We had only been there for a few minutes when the door flew open and three guards surged into the room. One held a long bamboo pole. They all looked George up and down in disgust as he maintained his vacant stare while water streamed down his naked body.

The guard with the pole snapped at Larry and me, "Wash quickly! Go back to your room!"

We rinsed the suds off, our eyes never leaving George. Larry and I didn't have to exchange words to know what we were thinking: there was no way we would leave George alone with these guys if we could help it.

The second guard pushed George, "Put clothes on!"

George's only move was to cross both his hands over his crotch.

The third guard shoved George again, harder this time. "Put clothes on!"

George dropped to the floor and curled into a ball.

The pole-wielding guard barked at Larry and me with increasing fury, "Hurry up! Finish washing and go back to your room!"

We turned off the water and grabbed our clothes off the bench. Larry reached for George's clothes and gave me a questioning look. I glanced at the guards, saw they weren't looking, and nodded at Larry. He tucked George's clothes under his. I knew what he was thinking: surely they wouldn't cart George away without his pants. They would have to come get them. Maybe that would buy us time. Time for what, I had no idea.

"Get out!" screamed the pole-bearing guard.

We scurried outside. Nobody followed us, so we lingered a moment, putting on our pants and gawking through the open door as the guards tied George to the bamboo pole, naked, dripping, and lolling. Two of the guards lifted the pole to their shoulders and began to carry him out like hunters hauling a tiger away after the kill. We trotted back to our room before they could turn their attention to us again.

Nobody locked us in our room, but after what had happened to the escapees we weren't about to try running away. We sat and waited, dripping, Larry clutching McSwain's clothes. Sure enough, about ten minutes later, one of the three guards stepped through our open doorway and said, "Give us McSwain's pants." Larry thrust out one hand, McSwain's pants dangling from it like a flag of truce. The guard walked out and locked the door behind him.

Larry and I stared at each other, and for once I could see that he was scared.

"Shit," I said.

"Yeah."

That was the last we heard of George for two weeks.

Larry and I alone together was always a tense situation, but this time most of our tension had a common cause: concern for George McSwain. That first evening, two guards opened our door and asked us to gather George's mat, long pants, long shirt, and his cup. They waited outside, so I whispered, "Put his socks in there, for the mosquitoes." I was sure he must be in the same building where they had kept me. Larry stuffed George's socks up a shirtsleeve. He then grabbed George's mosquito net and we both tried to push it up the other sleeve, but it made an obvious bulge, so we gave up. The socks would have to do.

A couple of days later, I was called to quiz.

At my quiz, I sat on a stool, while two guards stood behind me and GTO stood in front of me.

"Tell us about Bell stealing the book," GTO said.

"Bell didn't steal any book."

GTO's stolid expression didn't change. "You must write a letter to the camp commander explaining how Bell stole the book."

"I don't know anything about it."

That went over like a lead balloon. "McSwain told us everything. Now you will be punished for lying."

The two guards jerked me to my feet and for a moment I thought they might put me in the ropes again, but then they yanked my trousers down around my ankles. This was new. They pushed me onto the concrete floor until I was lying belly-down, my naked buttocks tensed for Lord knew what. Two more guards stepped into the room and stood on either side of me, each holding a long rubber fan belt. My survival instinct took over, and my body alternated between trying to rise from the grip of the two men who held me down and trying to sink into the concrete floor. One of the guards hauled back his fan belt and whipped it against my butt. I heard a sharp *crack* and felt a sting that made my flesh feel like a seared steak on a grill.

"Why did Bell steal the dictionary we found in your room?" GTO sounded almost bored.

"What dictionary?" I didn't dare say more. I had no way to know if McSwain had really talked.

Crack! came a second lash from the guard on the other side of me. My rear lit up with searing pain again, as if the fan belt were a hot knife.

"Did Bell steal the Vietnamese-English dictionary and my book?"

"No!" I didn't want to be responsible for bringing this same punishment down on my roommates, but how much could I take?

Crack! came a third lash. The pain intensity went up, nearly blinding me.

GTO spoke again, but I couldn't separate the sounds into words, could only anticipate the next lash.

Out of the corner of my eye I saw a guard move to lash me again. I struggled against the two guards holding me until I was running in place with my belly on the floor. "Okay! Okay! I'll write that you say Bell stole the book."

The guards released me. I pulled up my pants, easing them over my raw behind. I pulled up my stool to GTO's desk, where a pencil and blank paper waited for my confession. I wrote two paragraphs on why Larry stole the book:

Larry Bell had good intentions when he stole two books. We have been denied literature to read. It was nice to have a Vietnamese-English dictionary so we can learn a foreign language while we are incarcerated.

He only took the other book because we wanted something to read. Larry was not using the books to deceive the Vietnamese. I am sorry we read your books. I will not break the rules again.

GTO picked up the paper and read it slowly while the hot, stinging sensation continued to pulse up and down my backside.

I soon returned to my room, where The Boy Scout was pacing. "What happened?"

I wasn't about to lie to a fellow prisoner, no matter the consequences. I faced him. "They said McSwain told them you took the dictionary, and they asked me to confess. I refused of course, over and over. Then they hit me with a fan belt till I couldn't take it anymore. So I told them you took the books."

"Oh Bob." He shook his head.

"I wrote down that you only did it because we were bored and wanted something to read."

"Forget it." His tone indicated that he forgave me.

I wanted to say, Who do you think you are, that you're the one who should forgive me? I just took a bunch of lashes for you even though you're the one who took the books without consulting me about whether I wanted any part of it. I said nothing, but surely he could see I was steaming with rage.

"How many hits did you take?" Larry asked. I had known he would ask, hadn't I?

"I lost count after three," I said.

"Show me," he said.

"Give me a break, Larry."

"I just want to see how bad it is." He spoke as if he might merely be concerned.

I turned and yanked my trousers down.

He bent slightly for a closer look. "Yep. There are three of 'em," he said as if that settled something for him.

The next morning the guards took Larry to quiz. I had gotten over the worst of my anger at him, and now I was worried that after what they'd done to me they must be skinning Larry alive. After all, he was the thief, not me.

He returned about half an hour later, walking a little funny but not looking all that much the worse for wear.

He didn't wait for me to ask. "I denied it, and they gave me three fan belt lashes, and I still denied it, and they sent me back to my room." He shrugged as if it were no big deal.

What do you know? So maybe if I had denied it they would have dropped it after three lashes. Or maybe they wouldn't have. Maybe it all depended on their mood. There was no way to know.

I never claimed to be the fiercest of resisters. Still, during my time as a prisoner, I gave no propaganda, saw no delegations, went to no Christmas parties, and went on no trips to downtown Hanoi. I later learned from Air Force Brigadier General Robinson Risner that 95 percent of the American prisoners in North Vietnam did give propaganda to the Vietnamese. Either the researchers who unearthed that statistic had a different definition of propaganda than I or I was a much stronger resister than I thought, perhaps a little of both. I did not know back then that the prisoners who resisted hardest were typically forced into giving the most propaganda.

After that, it was a quiet two weeks for Larry and me, waiting for news of George, resisting the urge to tell each other off or punch each other out. The other prisoners weren't communicating much with us, thanks not only to the recent failed escape attempt and the false confessions about our involvement in that, but also thanks to George's fake insanity. For the moment, our room was trouble and nobody wanted that. So The Boy Scout and I sat in an information blackout, until one day the door opened and George himself limped in.

Unlike me, George had a big fat grin on his face before he turned around to bare his butt cheeks for Larry and me. His ass was a psychedelic Etch-a-Sketch of blacks, blues, and yellows. "They lashed me at least twenty times with a fan belt. I lost count after that. Guess you could say that stopped my insanity after, oh, about three hours."

Each of us always kept a lot to ourselves after quiz. Still, George admitted he had begged, "Don't beat me! Don't beat me! Don't beat me!" until he finally gave up any pretense of being crazy and agreed to talk to the interrogator if they would just stop whipping him.

He said the Naval Intelligence Officer pulled up a chair for the dear-old-Dad routine. I imagined George must have remained on the floor, unable to sit up for that particular heart-to-heart chat.

"Why have you done this?" the officer asked, referring to George's fake madness. "You were ordered to do this?"

"No, I just had a bad attitude. I just did it."

The officer told him that, after the escape attempt, they had interrogated the prisoners involved, and that's when they had uncovered George's game. "They told us about you, that you are not insane, that it was make-believe. We know about all the communications." The officer proceeded to tell him the nicknames we Americans had for different buildings, which rooms in those buildings communicated with each other, how they communicated, what kinds of messages they passed, and who the senior officers were. All of his information was spot-on. He tried to get George to admit that his behavior was a result of our communication system, a message passed from a senior officer to us. George held his ground, insisting he acted on his own against the advice of senior officers like Kasler. Which, of course, he had.

The biggest surprise of George's return was not his psychedelic rear-end but his huge grin, followed by quiet laughter, which he surely would have set booming across the prison if he hadn't wanted to spare us any more trouble.

He shook his head: "It was a hell of a spanking!"

Larry and I chuckled at his audacity.

"I can't believe you held out for twenty lashes," I said.

"If you can't stand the heat, get out of the kitchen!" George joked, but there was no rancor in it.

I couldn't help but like George, even if we didn't always get along. He was the real deal.

A few days later, July 28, 1969, the guards shuffled many of The Zoo's prisoners to new cells. George, Larry, and I were scattered across the camp. I didn't even hear about them through The Zoo's grapevine—although the grapevine did go on. The North

Vietnamese might beat the crazy out of George, but they'd have to kill us all to put a stop to our fellowship.

18

EXERCISE

Off I went to yet another prison cell, with three new roommates instead of two, in a small building near the front gate, known as The Garage. Even though my new room had one extra body, it had only one extra square foot of space. Four guys in a fourteen-by-fourteen room: if they were harder to live with than Larry and George, I would be in trouble. My new cellmates were three Air Force pilots: Lieutenant Terry Boyer, a business major from Northern California who liked archery; Joe Milligan, a New Jersey farm-boy; and Howard Hill, an Air Force Academy graduate from the border town of El Paso, Texas.

I was on edge that summer of 1969, wondering if the recent rough treatment I had received from the North Vietnamese after the escape attempt was a signal of things to come. It wasn't. In fact, over the next few months our treatment improved even more than it had in the months leading to the escape.

The guards allowed us outside for an hour or two each day, mostly to do cleanup, maintenance, and improvements around the camp. It beat sitting around and stewing together in the heat until we were ready to punch each other just for looking sideways. We made a new walkway, built a new camp road, leveled a foxhole and an old pigsty, and put up a volleyball court. We were handed hoes, pitchforks, picks, shovels, and even a wheelbarrow. With tools in hand, I felt more like a man again rather than just a prisoner.

The guards also let us outside to play volleyball for twenty minutes a day. Punishments almost stopped altogether. Other than being stuck in a cramped cell with three other prisoners for twenty-two to twenty-three hours a day, life seemed tolerable again.

I noticed something different about Joe, Terry, and Howard, compared to my previous roommates. Larry Bell had been careful not to so much as smile when a guard was in the vicinity, lest someone think he was cooperating or failing to resist the enemy, but my new roommates didn't seem concerned about guarding their behavior to that extent. They smiled at the guards, said hello, talked to them, even shared a joke now and then. It wasn't as if my new roommates and the guards did any special favors for one another or became buddies. It was more like a casual relationship between coworkers, except that our job was being prisoners and theirs was being guards.

After so much time with Larry and George, at first it made me uncomfortable.

One time I saw my new roommates laughing with a guard and said, "Geez, you guys, it's not going to look good if the other guys see you sitting here talking and laughing it up with the enemy."

Terry put me in my place with a smile. "That's the difference between you and me. I'm not worried about what other people think."

Another time, I made an offhand complaint to Joe about the enemy "keeping us cooped up most of the day."

Joe shrugged. "What do you expect? You're in jail."

On one hand, they made me feel stupid for the things I said. On the other hand, their attitude was enlightening. Larry used to see people coming back from quiz carrying a sandwich or an orange, or talking to a guard and smiling, and he would gasp and give me the suspicious *Wonder what he's up to* routine. My new roommates made it clear they didn't like being kept prisoner, but they also suggested we might as well try to get along with our keepers.

Still, it was only a matter of time before one roommate or another became the new itch under my skin. It was Howie.

At first, I felt sympathetic toward Howie. He had a huge heart, but it seemed as if Terry found him a little too good to be true. The two had been roommates for a while and had clearly spent some time working up a steaming case of mutual irritation. I didn't know exactly why. In theory, I accepted that prisoners with different personalities forced together by chance were bound to drive each other up a wall. But every cellmate was annoying in his own way, and knowing that it couldn't be helped—didn't help.

At 26, Howie was just a few months older than me, and we were both married. Based on that common thread we hit it off for a brief honeymoon. The resemblance ended there. He was a brain, a true believer in the American dream, and a tone-deaf but enthusiastic singer. Sometimes, he would tell a joke so quickly he would stumble on the punch line, which was kind of endearing. Terry would crack, "Typical Hill." At first I defended Howie. But after a couple of months locked up together, I found it easier to see him Terry's way.

As with most roommate issues, our biggest problem sprang from the tiniest detail: Howie was an exercise fanatic. I wanted to support him, but it was the height of summer when I moved in, with temperatures as high as 120 degrees, and Howie's exercise impacted the entire room. This kid from the southwestern desert was accustomed to heat, and no matter how much the three of us scowled at him, our disapproval was lost on him.

"The heat doesn't really bother me," he said. "I love summer!"

Meanwhile, I was a boy from the chilly Northeast who spent half my time in Vietnam with a heat rash. If it was especially hot, I had to be careful when I drank liquid—water, soup, tea, hot, cold—because if I drank more than a sip I would get prickly heat. It felt as if I were a voodoo doll and someone was sticking me with pins all over my body. I tried to keep a water bucket nearby to splash on myself and knock the prickles down, but it didn't help much.

Howie said, "Aw, come on, Bob, it's not that bad."

Howie was smart, but in his cabin-fever desperation to keep moving, he overlooked the thermodynamics involved in exercising in a room full of people. Put an extra person in a room and the temperature goes up. Have that person exercise, and that guy gets hotter. When he gets hot, everyone else gets hot. Howie's jumping jacks, toe-touches, and stretches only took a half hour a day, but it also took up half the walkable space in the room, relegating the rest of us to our beds. Oddly, Terry didn't complain. Maybe he thought it was a small price to pay to give a comrade a half hour of freedom of movement.

At the time, I only saw that half hour as a loss of freedom to myself.

Maybe that's because Howie's bed was next to mine, so he did most of his exercising in the space between our beds. His frenetic movements flung moisture all over me. What's more, when he did arm and leg exercises he often smacked and kicked me—not on purpose, but that didn't make it less annoying.

"Sorry," he'd say, breathless, but keep right on going.

After a month or so, I got up the nerve to say, "Knock it off, Howie!"

"Sorry," he repeated, and kept right on going.

One day, a bee got trapped in our room. Howie flipped out, flapping and swatting at it, trying to kill it or chase it out of the room—it was hard to tell which.

"Just leave him alone and he'll leave you alone," I said.

"No way, man! What if he stings me?"

Howie only managed to chase the bee to my bed, where it landed on my arm and stung me. The first bee sting of my life.

"Ouch! Didn't I tell you to leave it alone? But you just had to piss it off."

"Typical Hill," Terry said.

"Sorry," Howie said.

I did not hate Howie. But the heat made the worst of an uncivilized situation.

I was still grappling with the heat when September 2, 1969 arrived, the day Ho Chi Min died, twenty-four years to the day since his declaration of independence from France. The camp fell into such silent stillness, I was struck again by the realization that the Vietnamese leader whom we POWs saw as an embodiment of evil was a beloved hero to our jailers. I had never seen our guards look so somber, as they warned us in firm voices to be humble and not celebrate the death of Uncle Ho. My new roommates and I honored their request.

A couple of days later, the camp's food improved again. In addition to our two main meals, the Vietnamese started to serve us a tiny breakfast: either sweet hot milk or sweet toast.

One day a guard brought a hard-boiled egg for each of us, the first egg I had seen since my capture two years earlier. My roommates peeled their eggs and dug their teeth in with little groans of pleasure. My mouth watered as I cracked mine against the edge of my wood pallet. Then I dug a thumbnail in, and as I peeled the shell, I smelled the vilest odor of my life. My egg was rotten. My roommates giggled like schoolboys as I shoved the foul egg into our shit bucket and slammed the lid, trying not to vomit. I knew the laughter was good-natured, but I was still disappointed to miss out on such a rare treat. I never saw another egg during my time in prison.

The weather cooled off as the days marched toward winter, and for a time the heat that Howie added to the room didn't annoy me. Poor Howie was the only one of us who hated the cold. He would curl up in the corner of his bed, wrapped in his blanket and shivering.

"The cold doesn't really bother me," I would say. "I just love winter!"

He looked puzzled, unable to get how 60-degree temperatures might be a relief.

Just before Christmas, the guards let all the prisoners in the camp sit together to watch a couple of movies. We had gathered for movies before, but in the past the guards led the men from their cells to an already darkened room and made each group of cellmates sit separately from the other groups, some ten feet between each, no talking allowed. This time we gathered in a well-lit room, crowded shoulder-to-shoulder, with sixty to seventy-five fellow Americans. I don't know which struck me more: the feeling of brotherhood with fellow prisoners or of being overwhelmed by a crowd after so much time in a cell. The guards instructed us not to talk, but it was impossible to resist, and they ignored the infractions so long as we kept our voices low.

The first film was standard military propaganda, maybe thirty minutes of black-and-white footage showing brave North Vietnamese carrying supplies to the front line and winning battles, Americans killing defenseless villagers, and general rah-rah patriotism for the North. Or so it seemed; the soundtrack was in Vietnamese, which almost none of us understood. Maybe they hoped the visuals alone would be enough to convince us of their communist superiority.

The second movie was not typical propaganda, and though I didn't understand it I enjoyed it. It was about a troupe of Vietnamese puppeteers putting on shows in battle-torn areas. For about forty-five minutes, we watched puppeteers making puppets, practicing, and performing in the woods for audiences of children. At one point during a puppet show, an air raid came and they covered everything in camouflage. Then the raid ended, they unveiled the set, and continued as if "the show must go on." It was the first film the Vietnamese showed us that was not about the war but about their culture. I couldn't help but stare at the beautiful women and children. We prisoners had little or no contact with women aside from the women who made our food or an occasional guard's girlfriend who stopped by to peek at us. As for children, I hadn't spotted one in nearly two years, not since The Plantation, where the commander had a little girl I sometimes saw playing far across the compound. I listened to the film's music and singing, and watched the dancing, marveling that somewhere in the world families were still celebrating life. It was a nice change from their usual films of Americans bombing women, children, and old folks.

Christmas wasn't awful. It was the first time I was allowed to send a holiday card home to my family. The Vietnamese brought us several cards to choose from, with traditional Christmas images like evergreen trees, angels, and snowmen.

Come February 1970, they started allowing us to write home once a month. I was eager to let my family know I was all right and to ask for news from home. Still, I found myself censoring a lot of my words, not only because I knew the Vietnamese would look at everything I wrote before mailing it, but also because I didn't want my family to worry. Even though I was healthy and not suffering much, if I talked about my longing for home or the irritations and constraints of prison life it might upset them. So I kept the details lighthearted: the men I lived with, chores and activities I enjoyed, my health, my hopes of coming home, and my gratitude for the packages my mom sent.

My roommates and I were now allowed to receive packages from home every other month. A few months earlier, my mom had sent the biggest package I'd received so far, stuffed with such prizes as a sweater, long underwear, and socks for the cool winter; a couple of containers of Vicks Vaporub; toothpaste; pistachios; Carnation Instant Breakfast; instant iced tea; and bacon bits and other goodies I could mix with my rice. I couldn't see my way past saying much more than, "Thank you for sending such great stuff, Mom. You're the best!" Any more than that and I could imagine her reading between the lines and fretting.

Monthly packages from home became a bright spot in our otherwise dull days. My roommates and I were most excited to receive all-cotton t-shirts, tins of Vienna sausages, and hard candies. The cotton shirts were a relief in summer because they would get wet with our sweat and then the sweat would evaporate and cool us. I suffered a lot less from prickly heat that summer.

The Vienna sausages tasted of home, though we all agreed they left an unpleasant aftertaste in our mouths, maybe from the all-American fats and preservatives that were missing from the bland but natural meals served by our captors.

Joe Milligan received the best packages. His family sent Nabisco hard candies. We loved them because you could set one on your tongue and, oh God, let it linger there forever, unlike chocolate candies, which would arrive slightly melted, slip down your throat, and become a memory before you knew it. My new roommates and I created our own little ritual about those hard candies. No matter who received them we always shared, limited them to just one per person after each meal, and all ate them the same way. We would lie on our backs on our pallets, drop a hard candy into our mouths, and talk about life while the flavor flowed over our tongues.

One day after lunch we went through our hard-candy ritual, and when I started to talk I accidentally swallowed my piece whole. It got stuck in my esophagus. Not only was it a drag to lose my candy, it hurt like hell. I coughed and hacked and massaged my throat while the guys laughed at me.

When the pain subsided, I asked for another candy.

"No no no," Joe said. "The rule is only one candy per person per meal."

"Yeah, but I didn't mean to swallow it, and now I'm stuck without one."

"That's not our fault, so why should we get shorted?" Terry said.

"Them's the breaks," Howie said.

"That's cold." I kept my voice light, though it was hard not to think about how, back home, I could have simply gone to the store anytime and picked up more.

Around Easter, I finally discovered that my brother was alive, when I received a package from him for the first time. In it, he had included a photo of himself standing next to Dad. In the photo, Rich was wearing civilian clothes and bushy pork-chop sideburns. I thought the sideburns looked strange, but I was grateful to see my brother alive and in one piece.

There was way too much time in prison to just lie there and think. When I thought of home, mostly I thought about Pat, who tethered me to the hope of a future. But I thought about the past too, and about my family, and when I thought about Rich sometimes I felt guilty. Maybe all big brothers feel that way now and then, but that didn't make it easier. I was the first-born, and as

I grew older I recognized that sometimes I was treated like the favored son. When I broke the rules, Dad hit me, sure, but when Rich broke the rules, Dad hit him harder. When I failed, Dad told me off, but he shouted louder at Rich. When I came home with good grades, Dad expected the same from Rich even though he was three years younger. It would be years before the family realized that Rich was a genius. He was probably bored in school. If he was ever treated like second fiddle, he didn't complain. I was his big brother, and in his own easygoing way he looked up to me.

When I used to go home on leave, Dad would ask me to wear my uniform when we had visitors or went out. Sometimes it made me uncomfortable, like he was using me to show off, but he was my dad and I wanted to please him. I suppose Rich did too.

Although our dad was an American, he had joined the Royal Canadian Air Force during the Great Depression because he saw becoming a pilot as his ticket out of poverty. That was where he met my mom, who was a nurse in the RCAF. After the Japanese bombed Pearl Harbor, Dad joined the U.S. Army Air Corps. He flew The Hump, one of World War Two's most dangerous assignments. The Hump was the route transport planes flew over the Himalayas to deliver critical supplies to allies in the China-Burma-India Theatre. On some missions, as many as 50 percent of the planes would crash along that five-hundred-thirty-mile passage. Nearly a thousand men and six hundred transport planes were lost. Some are still there.

It was my father who first taught Rich and me to fly small planes. My brother and I liked it in the sky, so I suppose it was natural for us to see flying as our way to escape what we thought of as limited options.

Even though my twenty-three-year-old brother had those crazy sideburns, I could still picture him as a smooth-cheeked three-year-old, sitting next to me at the dinner table, where Dad sometimes gave us a nickel Hershey's chocolate bar to split for dessert. I wanted the whole thing, so I would make a deal with Rich, something like, "Let's play cards and whoever wins can have the whole Hershey Bar!" I was three years older, so of course I won. He never objected, just sat there with tears in his eyes watching me stuff chocolate in my mouth.

I never knew how much I loved my brother until I got locked up in a foreign prison.

Although I was relieved that he was alive, or at least that he had been alive when he sent the photo, his care package raised other questions that depressed me: Why has my wife stopped sending packages? Why doesn't she write me any letters? Was she in a car accident? Has she given up on me?

Meanwhile, Howie kept getting packages from his wife. His wife, who was still waiting for him, who hadn't moved on or died.

At times like that, I felt so isolated I looked at my roommates and thought, I'm not one of you. All three were Air Force, not Navy. I was a front-seat pilot, while they were GIBs, or "Guys In Back." In the Navy, we didn't have two pilots per plane, just one. The Air Force liked two pilots on board, so if the Air Force needed a guy to be a back-seater then he was a back-seater. GIBS were pilots too. Even so, it struck me that they were different from me.

The longer I spent in prison, the more I felt I was falling behind. I had gotten into the flight program with two years of college instead of four, so I had expected to get out of the Navy with a two-year jump on my peers. I was supposed to get out in 1968. Now it was 1970 and I was stuck in a foreign prison, losing my place in line at the airlines, falling behind in the seniority I could have been building, losing a little of my future each day. I had no college degree. I only knew how to fly. And there Howie, Joe, and Terry were, not even due to get out until later that year or next, and they were catching up to me. Not that I held it against them, but they were reminders of my private race to get on with my life, a race I feared I was losing.

As a POW with too much time on my hands, I did a lot of wishing. I wished for the airlines. I wished for home. I wished for my wife. I remembered holding hands with Pat on a sunny day in San Francisco, and walking into Sak's Fifth Avenue just for the fun of it. I told her maybe I could buy her something small as a treat. Pat found the tiniest treasure imaginable, a little jeweled money-purse, and we asked a saleslady to pull it out of the case for a closer look. We just wanted to check the price. We both saw the $500 tag and knew there was no way we could afford it, but neither of us blinked.

Not wanting to embarrass myself, I turned to my wife and said, "Pat, are you sure this is big enough?"

"No, I don't think so. I need something that will hold more than just change."

"Thank you," I told the saleslady. "We need to look around some more."

We pretended to shop a moment longer, then strolled out, cool as you please. When we reached the sidewalk, we both leaned on each other and exploded with laughter.

Sometimes the memory of Pat's face lit up with laughter kept me from just lying down on my hard wooden pallet and giving up. Then again, sometimes the effort of remembering her face at all had the opposite effect.

"Cheer up, buddy," Howie said. "I'm sure she'll send you something soon."

"Give it a rest, Howie. You don't know what you're talking about."

Howie wasn't the type to give it a rest. He kept right on with his relentless optimism, and his daily exercises, until the 100-degree-plus heat set in again.

"Howie, why don't you knock off your exercise! It's making the room hotter. How many times do I have to tell you I'm sensitive to the heat?"

"You must be used to the heat by now. I'll be done in a minute."

I stood, stepped into the space between beds, and faced him. Joe and Terry sat up on their pallets and stared at this unexpected break in the monotony, maybe waiting to see if this was going to be entertaining or dangerous. "Howie, you're pissing me off! Be cool. Use some tact."

"You're not the only person in here, Bob."

"That's my point."

We paced, two animals defending a territory that wasn't big enough for both of them.

Howie said, "You know, maybe if you give the bitching a rest, you'll feel better."

That only made me angrier. "Howard, give it a break!"

He knit his brows and snickered, as if he couldn't believe I was taking myself so seriously. "Oh, stop crying." Then he turned his back on me, still chuckling.

That was when I lost it. He had sixty pounds on me, mostly muscle, but I wasn't thinking of that. I barely knew what I was doing when I reached for the first handy object, our black tin bucket full of crap and pee. I picked the shit bucket up by the handle and, while his back was turned, swung it at his head as hard as I could. A hollow *Bong!* reverberated through the room. I felt a split-second of satisfaction that I finally had his attention, followed by fear that I might have just killed him.

Howie groaned and staggered into Terry's bed.

Terry and Joe were both shouting. One of them said, "Holy shit!" as they both jumped on me and dragged me clear of Howie. That's when it occurred to me how big Howie was, but that didn't make me back down. Just the opposite. I figured I had to make a show of it now because Howie might be angry enough to beat the crap out of me. I have to scare him, I thought, make him think I've gone off the deep end.

I screamed, "Get him away from me! If I ever get the chance, I'll kill him in his sleep!"

Joe was a wrestler who weighed about 170, so it was nothing for him to shove my 110 underfed pounds to the concrete floor, where he sat on me and shouted at Howie, "Get back, Hill! Stay in the corner!"

Terry hovered above us, the voice of reason. "Be careful, Joe. Watch his head."

And Howie? He stayed in the corner, repeating, "I'm sorry, Bob. I'm sorry, Bob." He sounded truly worried that he had caused me mental damage.

Despite that, I was still scared of what he might do. We were far from any sort of law-and-order American style, in a war-torn country, prisoners of the enemy, and all I could think was that it was possible Howie could beat me to a pulp and get away with it. How could a little guy stand a chance against a big guy in the middle of a prison? So I struggled and yelled and acted crazy enough to kill.

Joe tightened his arms around mine in a bear hug, pressing me flat to the ground, and said in a tone so reasonable it would have sounded funny under other circumstances, "Oh shut up."

Unable to budge, I went limp. "Okay, Joe. I'm okay. Just let me up."

He released me, I scrambled to my feet, and in the sudden quiet I became aware of one of the guys next door tapping on the wall: *shave-and-a-haircut*. Then I heard a muffled voice vibrating through a tin cup pressed to the wall: "What's going on over there?"

Terry Boyer grabbed his cup, pressed it to the wall, and talked into it. "Wideman berserk. Call the guards!"

That wasn't the kind of news I wanted making the rounds. I wanted to scare Howie, not to be known as the camp nut-job. So I went to the wall and said, "I didn't go berserk. I just hit Hill."

It wasn't long before two guards appeared at the door. "Wideman, roll up," one of them said.

I didn't object. In fact, the order came as a relief. I didn't want to be in that room another minute. I rolled up my stuff and followed them out without looking back.

This time I spent eleven days in isolation. It gave me a chance to cool off, but if it was meant as punishment it didn't work. As usual, it was a blessed reprieve to sit all alone in a room, my only communication a few messages carved into the bottom of my dishes.

I was called in for quiz. The Baboon was waiting. He asked me what happened. Though he wasn't the handsomest interrogator in the bunch, he wasn't the scariest either. Under the circumstances, I figured it made sense to tell the truth, apologize, and move on.

"Hill and I have a personality conflict. We don't get along."

The Baboon looked disgusted. "Hill is mad at you. Your roommates are mad at you. Hill is hurt. He has a bump on his head."

"Oh shit." I cast my eyes at the ground, honestly ashamed. Now that I was out of the situation, the whole thing looked different. I felt like an ass for nailing Howie—with a metal shit can of all things. It was just a silly argument, and I had allowed it to reduce me to brute animal instinct. "I'm sorry. I didn't mean to do it. It was hot. I lost my temper. It won't happen again."

The Baboon and I agreed it would be better if I had a new roommate, but he gave me a condition. He said that first I had to finish a painting I had been working on, a project I'd started before I went into solitary. As part of our improved treatment, all of us had been given papers, paints, paintbrushes, and American postage stamps, so we could paint copies of the stamps. Mine had a picture of President Franklin D. Roosevelt. I figured The Baboon was trying to take advantage of my mistake so he could create propaganda: Look at the pretty paintings the happy American prisoners make! See how well we treat them. We even let them paint. I didn't risk outright defiance, but I did a sloppy job on the painting.

That pissed off The Baboon. He said I would have to move back in with Hill. But first, he demanded I write a letter of apology and ask Hill if it would be okay if I returned. At first, I didn't have a problem with that, since it was not propaganda. And I really was sorry.

The letter started off simply enough: *Dear Howard, I'm sorry I hit you...*Then I started thinking maybe the North Vietnamese could use it for propaganda after all, showing how brutal the evil Americans were or something like that. So in the last sentence I threw in something that would make it too embarrassing for them to use: *And I want to live with you, Howard, because I love you so fucking much I could shit. Love, Bob.*

The interrogator took my letter to the camp commander and returned a short time later. "The camp commander does not think this is funny. I want you to change it."

I wrote another one and tried to keep it as short as possible, hoping that would make it harder to twist my words into any kind of propaganda. After that one, I was sent back to the room I shared with Howie, Terry, and Joe.

When the guards let me into the room, Howie jumped from his pallet and said, "I'm not staying with this crazy guy! No way! You have to put me somewhere else. I'm scared." The guards stared at him as if *he* were the lunatic, and led him out.

It wasn't long before the guards brought him back, silent and uncomplaining.

When they left, Howie turned to me and said, "I didn't do that because of you. It was just an issue between me and them." They had quizzed him too. It hadn't gone well.

I didn't ask for an explanation. I just looked him as straight on as I could manage and said, "I'm the one who should apologize, Howie. I guess I made an ass out of myself. I know apologies don't do any good, but I don't know any other way. I hope you'll forgive me." I held out my hand. I was relieved when he shook it.

It could have been worse. A few months later, I would meet Air Force pilot Ed Mechenbier, who had gotten into a fistfight with his roommate, Kevin McManus. Kevin had also been his GIB (Guy in Back) when their F-4 was shot down, and it seemed he blamed Ed for the shoot-down. He slugged Ed in the mouth and broke his tooth. After that, the gum became infected. Ed endured recurring pain from that abscess for years. We weren't the only ones. Every now and then, prisoners who just couldn't take any more took it out on each other.

I always had more of a problem getting along with other prisoners than with the Vietnamese. The Vietnamese and I had an unspoken understanding that they could get me to say or do anything simply by applying pain in a specific way. Between us prisoners there was no such understanding. Only the chaos of men who shared little in common but a longing for escape and home, and the uncertainty that either would ever come.

19

ORDERS

In July 1970, the North Vietnamese told a large chunk of prisoners throughout The Zoo to wrap up our belongings for a move to another prison. Spending two years in one place had not made it feel like home. Still, it was familiar. We didn't know why we were moving or where we were going. There was a chance things could be worse at the next place, though we hoped they'd be better. There was also a small chance we were moving because the war was over or, God help us, because the end of the war was nowhere in sight.

The Zoo was in the Hanoi suburbs and, although I was blindfolded for this move as usual, it was easy to tell by the decreasing noise that the new camp was farther outside the city. We arrived at a camp divided into four concrete-walled compounds. Each compound contained an identical long building. The guards took me to one of those buildings, which had seven large rooms, all much bigger than my room at The Zoo. I was one of seven prisoners in my room, the most roommates I had lived with so far. My new roommates and I did the math and figured there were about two hundred prisoners in this no-name camp.

We gaped at the large barred windows on both sides of our room. A few of us stood in the middle and closed our eyes to enjoy the relief of cross ventilation. The open space between our building and the prison wall was about fifty feet wide with a broad swath of fresh grass. The sight of that human-tamed greenery seemed unreal, like a child's crayon drawing of a yard. At one end of the building was a deep cistern where we would be allowed to bathe twice a day.

The guards let us out for two to three hours in the morning, the same in the afternoon. After three years of roasting indoors, I had forgotten what it was like to spend hours in the sun and air. One day, a bunch of Vietnamese men showed up at one end of the building with hammers, nails, and a ladder. The men spent a week building an open-sided shelter, a simple bamboo roof on four poles. They installed a Ping-Pong table underneath so we could play Ping-Pong when it rained.

Over time, we began seeing a regular doctor, who would visit the camp two or three times a year. He played the role of dentist too, sitting patients in his portable metal chair to pull teeth and treat abscesses. Despite the NVA and our families back home keeping us all supplied with toothbrushes and toothpaste, plenty of guys had one pain in the mouth or other, and without a full-time dentist they often suffered for long stretches. I felt lucky not to have any problems in that department.

By my standards, this new camp was a country club, low-budget maybe, but fairly easy living. It was clearly meant to house us long term—not a timeframe I wanted to dwell on, but it could have been worse. I believed this would be our last move until the journey home. Some prisoner or other nicknamed it Camp Faith, and the name caught on.

Our faith was not rewarded.

In the fall, an unmanned American drone flew over Camp Faith. Then again a day or two later. Then a day or two after that. We figured it was a reconnaissance plane and wondered what kind of mission it was prepping for. Might this be the operation that would end the war? No idea. But we thought the drone might be taking pictures of our camp. One day we were at the cistern washing up for chow when one passed. My roomies and I looked straight up and smiled to make sure the camera could get a clear shot of our faces.

One morning in November 1970 I woke to hear exciting news bouncing from roommate to roommate. Several had woken the night before to hear guns going off and planes flying overhead. "One plane buzzed right over our camp. Boy, was he moving!" someone said. I was amazed I had slept through the first air raid in more than two years. It seemed those early years had habituated me to sleep through almost anything.

The morning after the air raid, I saw a handful of high-ranking North Vietnamese officers walk into our compound and take a look around, shaking their heads in disgust. That same morning the guards told us to wrap up for yet another move.

I looked around the camp, and saw POWs in other buildings pulling down their clotheslines. The whole camp was packing it in. No doubt it had to do with the planes and guns from the night before. I wanted to hope that this meant the war was over and we were going home—I couldn't imagine any other reason they would move us out of a spanking new camp just four months after settling us in—but hope was hard.

Back into the city we went, some two hundred of us herded into the main prison in downtown Hanoi. The Vietnamese called it Hoa Lo, but it would become known among us prisoners, and later all of America, as The Hanoi Hilton. There, they put fifty of us into a large concrete room the size of a gym. Concrete beds lined either side of the room and a wide walkway went down the middle. One end of the room was separated into a concrete-walled shitter with two holes to squat over. Two holes for fifty people. No wonder the wall of the shitter was regularly plastered with flies.

I'd never been housed with so many men before. Still, the room did not seem crowded. In fact, I allowed myself a surge of optimism as I looked around: Maybe we really are getting ready to go home. Maybe they're putting us all together in one place so they can load us on trucks or a train or plane or something and send us to wherever the exchange point is.

It was lunchtime when we settled in. As we ate, we reoriented ourselves to yet another home that was not home. We were in a walled compound at the back of the camp, in a four-wing building that lined the perimeter. The building's several large rooms faced a central courtyard. Next door to us were about eighty South Vietnamese men, thirty more than we had in our room even though theirs was the same size. Another room held Vietnamese women exclusively, but the guards quickly set up bamboo screens to shield them from our sight.

The South Vietnamese men went outside the camp most days to do roadwork. In the evenings we established communication with them. It wasn't hard to pass notes to each other. Our doors were close together and each had a barred opening at the top and a gap at the bottom, making it easy to either kneel down or climb up and toss the notes across. Reading each other's messages was another matter. They only knew limited English, so the messages seemed cryptic, featuring the names of their commander and outfit and drawings of a battle: men in parachutes jumping from planes, helicopters flying over buildings, soldiers with guns, bodies. When we didn't get far with that, they wrote in Morse-code dots and dashes. We figured out they were trying to tell us about a battle the Americans had won.

Americans doing good work, one note said. *Vietnamese appreciate what Americans do.*

We soon learned through the wall-to-wall grapevine just what our South Vietnamese neighbors were trying to tell us. We had been relocated because the American military had raided a POW camp and attempted to free the prisoners. The Americans killed about twenty-five Vietnamese in that raid. The North Vietnamese were not happy.

It would take months for us to piece the details together, and the picture was never clear until after the war. While we had been living at Camp Faith, U.S. forces had attempted to rescue POWs from the Son Tay camp, the next camp north of Faith. Intelligence analysts reported that as many as fifty-five prisoners were there. The raid was a joint operation: an Air Force assault group flew fifty-six Army Special Forces troops to the camp under cover of darkness, while the Navy created a diversion by flying over Haiphong Harbor and dropping flares to simulate an attack. More than a hundred aircraft were involved.

Imagine the Green Berets' surprise when they found no prisoners in any of the cells! They reported "negative items" on the radio, boarded two helicopters, and withdrew. The whole operation took twenty-seven minutes. The Americans suffered only a few minor injuries.

The NVA had moved us back into the city to stymie future rescue attempts. Still, many prisoners were overjoyed about the Son Tay raid, seeing it as a signal that we would all be rescued soon. On the other hand, some feared the NVA would kill us if the Americans raided the camp. Only a handful of us whispered *my* speculation: that our government would never have attempted the rescue unless our side was desperate. The writing was on the wall: We were not winning this war.

In the days to come, we did an informal inventory of who was at the Hanoi Hilton: more than three hundred fifty of us. For the first time we were all in one place. In general, many of us were grateful we were together, in big rooms no less, where we could communicate better than ever before.

One thing I appreciated about the large number of men in one room: I spent more time with buddies of my own choosing. Kevin McManus, Ron Bliss, Ed Mechenbier, Guy Gruters, Jim Shively, Joe Crecca, and Tim Sullivan were an educated group of men, with common sense to boot, and I admired that. About a dozen of us created our own little school, with a handful of guys volunteering to take turns teaching. We gathered on four beds around the concrete abutments along the walls for a couple of classes a day. We were eager to use our time to better ourselves rather than just marking off the years. What else did we have to do?

Ron was an expert in world history, especially World War Two. Guy taught advanced math, including calculus, a subject I struggled with. Tim, a Catholic who graduated from Loyola, taught religion, philosophy, and sociology. Joe taught music appreciation. Kevin taught art appreciation and French. Ed taught German. (Ed also happened to be the pilot who still suffered from a broken tooth and abscessed gum after a long-ago fistfight with his cellmate.)

My new teachers and friends used their strong backgrounds in arts and humanities to tutor me in understanding that everything in our world is connected. I came to see that life's events, from family fights to wars, don't happen in a vacuum. Every action has an impact, creating ripples with the power to touch every aspect of our world: philosophy, law, economics, politics, science, and more. We all had strong opinions and didn't always agree, but our exchanges taught me a lot, especially a new appreciation for the value of education.

Possibly my favorite subject to learn from fellow POWs was strategy at chess and cards. Kevin had an old travel chess set his family sent him from back home, a wallet-sized job with pegs in it, while Galen Kramer was an expert at card games.

Galen came from a family of means. His father had started as a used car salesman but struck it rich in oil. Galen was six foot one with the rugged looks, charm, and confidence of Chuck Connors, the actor famous for playing *The Rifleman* on TV. Like many of us, Galen also had a beautiful wife waiting back home. For all his luck in life—up until his capture of course—he didn't have an ounce of pretension. He was the kind of easygoing guy everyone liked.

It was Galen who dared to say after the Son Tay raid, "You can bet we're not going home anytime soon." But then he shrugged, as if to say, No point crying about it. What good would it do?

Galen had asthma and every winter he came down with pneumonia or something like it, breathing as if he were underwater, coughing up phlegm, his roommates wondering when he would die. Yet he never complained. Roommates took turns staying up with him at night and rubbing his back while he struggled to breathe. But the Galen I remember was always giving us a self-satisfied look over the top of a fan of cards, knowing he was about to beat us at gin rummy or bridge.

Perhaps my greatest teacher in prison was Jim Shively. Jim was a talented F-105 Thunderchief pilot with a master's degree in international relations from Georgetown. He taught us political science and Russian. What a difference from The Boy Scout, whom I sometimes called "The Graduate" because he had a degree in electrical engineering. Where Jim considered the connections between world events and the motivations of world leaders, it seemed to me Larry bought the military's party line, accepting without question whatever his superiors said.

Jim and I often discussed the war. He said we had already lost, at the Tet Offensive in '68.

"No way!" I said. "I mean, sure, that was a setback, but America is a superpower. They're not going to let a setback stop them. The U.S. never quits until it wins."

He smiled. "Everybody loses sometime. We're due. And the Vietnamese will go down in history as defeating two colonial powers. They defeated the French, and they'll defeat us."

I could never win a debate with Jim, but I still didn't buy it. "I'll believe it when I see it."

Despite my protests, Jim had planted a seed of worry in me. He was a smart guy. What if he was right? I had bought into this war because I believed the Domino Theory: that if Vietnam fell to communism, more countries would follow. I feared the communists would target countries surrounding ours, isolating us and starving us economically. After all, wasn't that the strategy we used against our communist enemies like Russia, China, and Cuba?

It would take a lot more conversations with Jim to talk me down from placing so much confidence in any political dogma. He said that no government lasts forever. He explained that rebellion leads to power, power corrupts, and corruption leads to more rebellion.

I would soon get several firsthand lessons in what Jim meant.

I believe it was on New Year's Eve that one of our room's leaders decided it would be a gesture of good will to sing for the South Vietnamese prisoners next door. We all readily agreed. We gathered facing the door and sang holiday songs like "Jingle Bells" and "Silent Night," and patriotic tunes like "God Bless America" and "The Star Spangled Banner"—not to make a political statement but because these were the only songs we all knew. A few guys climbed to the top of the door or crouched at the bottom to smile and wave, and gave us a thumbs up that the South Vietnamese had gathered to listen. I felt united with my countrymen, as we used the universal language of music to tell our neighbors we were all in the same boat. Encouraged by our cheering audience, we cut loose, getting louder with each song.

The next day, the North Vietnamese tightened up on camp rules. A guard stood in our room and announced, "You make too much noise. No more singing in the rooms."

One of our leaders said, "Are you saying we can't sing during religious worship?"

"You may sing on Sunday, but only four people, and you must sing quiet." He explained that it was not a punishment but for our protection. "If civilians know you are in the city again, it will make our job of protecting you very difficult. Many Hanoi people want to kill you."

The next Sunday all the rooms held regular church services. As usual, about thirty people in our room attended. Also as usual, *all* of them sang, not just four of them as the guard had ordered. The only change was that they sang louder. The sound of some two hundred Americans singing religious and patriotic songs at the top of their lungs reverberated throughout the compound.

About fifteen of us listened in silence, including Jim Shively and me. I was still the nonbeliever my father had made me and I never attended services anyway. Even if I had, I'm pretty sure I would have followed the order to keep the noise down and save our skin. Jim and I exchanged glances that seemed to say, I know they want to be brave, but this is stupid. I tensed, wondering who would descend on us first: angry guards or a lynch mob?

I sighed with relief when it was only the guards who showed up, shouting, "Okay, no more singing!"

Lieutenant Jack Rollins said, "Just like those communists: no freedom of religion!"

I couldn't believe his reaction. I didn't love my captors, but in this case it seemed clear: they were using common sense to protect us, and themselves, from a Hanoi lynch mob, not suppressing our freedom.

After the guards ordered an end to the singing, a chant rose from one of the rooms, "We are Room One, who are you?" Then the next room took up the chant, "Hey, hey, in Room One. We are Room Two, who are you in the next room?" The next room chanted, "Room One and Room Two, we are Room Three, who are you?" And so on until the chant was bouncing all over camp.

The guards immediately rushed through the prison and ordered everyone to be quiet or face punishment. They accused our leadership of trying to incite a riot.

After the war, Air Force Lieutenant Colonel Robinson Risner would report that the chanting was spontaneous, that the leaders did not start it but did not stop it either. "It was a riot, but it was fun," he would say. Fun? Maybe. Unnerving? Definitely. I still thought it was a foolish move that could have gotten people killed.

Before the chanting riot, many of our leaders were living in one large room along with the rank and file. We're talking colonels and commanders—"The Heavies" as I thought of them. The Hanoi Hilton marked the first camp where so many of our imprisoned big shots were gathered in one place, and it presented an exceptional opportunity for them to collaborate. After the chanting riot, the NVA must have realized this was a potential problem. They soon split up The Heavies to prevent further insurrection.

The four colonels were sent to a couple of two-man rooms. The officers just below them, including Risner and Navy Commander James Stockdale, were sent to another batch of two- to four-man rooms. The colonels set up a pyramidal organization with themselves at the apex. At the time I saw this as logical. I couldn't foresee the inevitable problems of *prisoners* trying to maintain order with a formal military hierarchy.

Perhaps the biggest problem was that orders from the big four had to pass through middlemen, and several walls, before trickling down to the rest of us. The same happened moving in the opposite direction. Risner and Stockdale got caught in the middle of the childhood whispering game, Telephone: the message that went in one end never came out the same way on the other. This exacerbated a problem I had not really paid that much attention to before I was imprisoned: much of our military leadership had always been a little out of touch with the men they were supposed to lead. But now they were trying to enforce a hierarchy in which the various levels lived in forced isolation from one another, making it that much harder to maintain the smallest grasp on the reality of what life might be like for the rest of the camp.

In January, we were all eager for mail call, because that was when our Christmas packages rolled in. Several prisoners bobbed on the balls of their feet as we lined up outside a room where the guards had set up a desk. At the front of the line, Squirrel was waiting to help me. He was one of the friendlier turnkeys who rarely shouted or gave anyone a hard time. He searched through the line-up of packages behind him and brought back one of the biggest, maybe two and a half feet long by a foot and a half square. It was stamped *Cleveland, Ohio*, which told me it was a package from my wife.

Squirrel gave me a grin of anticipation as he set it on the desk. Then I opened it. I didn't see anything. I peered over the edge of the box. Way down in a corner I spotted a can of nuts that had been opened and was only two-thirds full, a little bag of milk chocolate bells wrapped in red and green Christmas foil, and a bag of hard watermelon candies. That was it. Most of the package was empty. I couldn't stop staring, as if it might have been a trick of the light.

I looked up at Squirrel, and saw my loss reflected on his face. The tears I refused to shed filled his eyes. We both knew the truth. The guards who checked the mail for contraband or secret notes from our government regularly skimmed the cream of our care packages for themselves. I usually took it in stride, though I didn't blame other prisoners for getting bent out of shape when we had so little to look forward to: "They chopped up my M&Ms into tiny pieces," or, "They opened up all my puddings and half-emptied them." But never had I seen a package so vacant.

For a moment, Squirrel and I said nothing, just stared together into the void of the box. What good would it do for him to apologize, or for me to complain? I knew he wasn't the one who had taken my things, and there was nothing he could do. Instead, I smiled, held up the package, and tried to look grateful. "Thank you. It's a present from my wife." I tried to take comfort in the knowledge that at least Pat was still thinking of me. I would never learn what she had sent.

That was the only time the guards stole something tangible from me. But pilfering guards were the least of my problems.

20

THE HEAVIES

In January of 1971, three and a half years after I became a prisoner, at the moment when most of us had high hopes of going home soon, our commanding officers decided it was time to lead. The first task they took upon themselves was to set up a two-tiered communication system: one open and intended for all to hear, the other coded and intended only for the commander of each room and his covert information team. With that in place, the first task they assigned themselves was to get an official headcount of the prisoners at the Hanoi Hilton and determine who the officers were in each room. This seemingly simple task showed their complicated communication system to be a comedy of errors.

The brass named Major Ray Dennis to lead our room's four-man comm team because he was higher ranking than most of us. Ray had been an Air Force bombardier navigator. The problem was: Ray was slow. Like a turtle. He walked slowly, talked slowly, even listened slowly, if you can imagine that. He wasn't the kind of guy people expected big things of. So being assigned to the comm team? Ray loved that job. Finally he was in on the action.

On his first day at the wall, he was working it, like James Bond on a mission: he craned his turtle neck at the window to make sure no guards were in sight, crept up to the wall, tipped his ear to listen to the tapped message, made exaggerated facial expressions so everyone watching would know this was important stuff, and carefully tapped a response. It took at least forty-five minutes. He craned his neck again to make sure no guards were looking, then dramatically waved all of us over.

"We've got communication with the guys next door," Ray stage-whispered.

Just tell us already, I thought.

"The senior man is Doug Clower. The line-up is..." We already knew most of the information he relayed. Everyone he listed was a guy we knew. We'd seen them coming in, doing their wash in the morning, waving out their windows. But Ray continued to list every name like this was the greatest moment of his life. Who knows? Maybe it was.

Though Ray wasn't the best of communicators, we soon understood why The Heavies had ordered a formal headcount: they had decided to create the 4th Allied POW Wing, right in downtown Hanoi, and to do that they needed to establish a chain of command.

The Heavies declared each room a squadron and appointed the senior man in each room as squadron commander. Each squadron commander then appointed an executive officer as his second-in-command, and also appointed a communications officer and a supply officer to serve as his staff. The commander divided the entire room into four flights and appointed four flight leaders to head up each flight.

There were plenty of junior officers who received no special post. I was one of them, and as such I was expected to keep my mouth shut, follow orders, and not make waves.

Lieutenant Commander Al Stafford became our room's squadron commander. He was thin, six feet tall, maybe 135 pounds. He had small, cold, blue-gray eyes, and the pupils were always pinpoints whether or not he was in the sun. My first memory of Al was the day I returned from my morning bath to find him surrounded by an audience of about half our room, some two-dozen guys staring as Al pulled an eighteen-inch worm from his throat, like a sword swallower in reverse. It was gross. Most of us POWs ended up with intestinal worms sooner or later, and a few of my roommates and I had found hideously long ones in our stools, but most of us didn't make a show of them. Al was the only guy I ever saw pull one from his mouth and hold it up, covered in slime, like some sort of trophy.

Over time, I would come to believe Stafford was sick—not sick as in disgusting, but sick as in having mental or emotional problems. He was one of those people who can't face the wee hours alone without going nuts, so he forced us to suffer with him, talking half the night away. If the room fell silent, he tossed random questions into the dark, anything to keep the conversation going. I learned why one night when I woke to the sound of screaming.

At first, I thought maybe Kramer was dying of an asthma attack, but that made no sense because asthma makes it hard to breathe and these convulsive shrieks were using plenty of oxygen. Someone whispered that it was Stafford: "Dude gets nightmares." I had heard that he once caught his wife in bed with another man, and that he once pulled a buddy out of the wreckage of a burning plane. It was hard to tell how many of the stories we heard about him were true because he had a reputation for making things up.

Certainly we had all seen things that could make it hard to sleep.

Up until the Hanoi Hilton, ranking officers had rarely called anything they asked us to do an "order." Now it was as if they wanted to make up for lost time. Leaders who gave themselves code names like Abe and Sky Blue invented a host of orders to cover every possible thing that might happen in the course of prison life, and made it clear that we could be disciplined, even court-martialed, for failing to obey. They came up with code words for every order. The codes were supposed to confuse the Vietnamese. Mostly they just confused us. There were so many orders and code names that each room had to form a small cadre to memorize them.

The move to reestablish discipline may sound reasonable, but it came after several years during which all of us had figured out everything on our own, from communications to interrogations to daily routines, with virtually no leadership from our senior officers—and we had handled ourselves in a way that I believed would make our country proud. Up to this point, the only real direction I had received from senior leadership had come from Colonel Risner. When prisoners had asked him how much punishment we should take before yielding, he had replied that we should take it as long as we could, but not to the point where it would cause permanent physical damage. I thought that was a sensible policy.

At first the new orders only codified what most of us had been doing all along. We were ordered: *If you are asked to give military information that you've already given, you will not give it again until you have reached significant pain.* Our leaders also insisted we press the Vietnamese to treat us according to the Geneva Convention. This didn't make sense to me because our treatment was already better than the Geneva Convention required.

During our first six months at the Hanoi Hilton, our leaders circulated several polls. One asked how many days the North Vietnamese had tortured us. That poll estimated the average number of "torture days" at twenty-five, not including time spent kneeling or standing in a corner with our hands in the air. I was amazed at how low the number was, considering how easy it is for humans to rationalize brutality during war. As hard as I racked my brain, I could only come up with sixteen "torture days" I had endured. I felt bad about that because, if the poll was correct, it meant that theoretically some other poor prisoner had to endure nine extra days of "torture" to make the twenty-five-day average.

The new rules did not stop at the subject of torture. If I had a medical problem, I could not simply ask the Vietnamese for medicine. I had to first tell my senior officer, or SRO, and he would determine whether I really needed it.

We had orders pertaining to escapes, rescues, communications, cover stories, and resisting the Vietnamese.

We memorized codes for resisting our guards. If there was a conflict between a guard and the prisoners, our SRO could order us to go to *Stand, Soldier, Squat,* or *Sing,* and the room was expected to go into the designated resistance posture. If the SRO called out, "*Stand,*" then we were supposed to stand rigid and ignore the guards, refusing to bow, talk to, or look at them. If he said, "*Soldier,*" then we were supposed to do almost the same thing, except now we were to give menacing stares directly to the guards—which usually meant looking down at them because even short guys like me were taller than most Vietnamese. The only time a resistance order of this sort went into effect was when the room next door went to *Squat,* which meant refusing to eat. They did that in protest of the Vietnamese ordering them to make coal balls. As for *Sing,* after our first, unofficial, version of that act of resistance we never tried it again.

About a month after The Heavies instituted the new comm system and codes, right about the time the comm teams were just getting the hang of it, we heard word that The Heavies were communicating with another room while, unnoticed by any of them, a guard sat in a corner between the buildings with a pad of paper and pencil, writing down everything they were saying. What they were talking about: the importance of maintaining standardized communication procedures so we would not get caught. That's right, the leadership had put us through the hardship of changing our comm procedures, which had been working just fine before they came along, and all it took to disrupt their new system was one guard armed with pencil and paper.

After that, The Heavies decided to solve the problem by making up a whole new list of code words for us to memorize.

That snafu didn't slow down the stream of new orders. The list kept growing from 1971 until the end of the war. There were rules about what the guards could give us. For example, we might be able to borrow guitars for recreation, but could not accept personal gifts like books. One order declared: *You will not read, or accept into your room, the book called* The New Soldier. Another order said: *You will not associate with Miller or Wilber.*

I didn't know what Ed Miller and Gene Wilber had done to bring that punishment down on their heads. The official communication to us masses was that they were "working for peace," which insinuated that they were working with the North Vietnamese, or at least not supporting the U.S. military, which as far as The Heavies were concerned amounted to the same thing.

Back in '69, the NVA had played a twenty-minute tape for us in which Ed Miller and Bob Schweitzer had made antiwar statements, things like: "Eisenhower said that eighty percent of the Vietnamese would have voted for Ho Chi Minh" and "These people just want self-determination and they're not getting it." The prisoners called it *The Bob and Ed Show*. I never knew if the NVA had tortured or threatened Bob and Ed to convince them to make the tape, or if they really believed what they were saying. Either was possible. Not everyone believed in the war. Whatever the reason, we weren't supposed to associate with Miller and Wilber—even though Wilber wasn't even on that tape. The Heavies declared

Bob Schweitzer was now on our side and running the show in his room, while Miller and Wilber were relieved of command.

When those sorts of orders came down, my moral indignation set in. I did understand that by joining the military I had given up some of my rights to freedom of speech and assembly. I realized that our leaders forbade us to read certain books or talk to certain people because they believed it was necessary to maintain order. But to me, many of the new rules went beyond maintaining order to the point of attempting to control our thoughts. The Heavies seemed determined to suppress any views that did not conform to their own. Ironic, given that we had come here to fight a communist dictatorship.

Even when our leaders weren't handing out orders, much of the communication they passed on was about coercion, control, and manipulation. The colonels passed down such advice as "The Five Do's." I remember two of the five: 1) *When you wake up in the morning, ask yourself, "What can I do to support my SRO today?"* and 2) *When you go to bed each night, ask yourself, "Have I done all I can to support my SRO?"*

It seemed to me that the only reason for changing to such a rigid command structure was because there was a chance we might all go home soon, and the senior officers wanted to tell their superiors they had accomplished something while they were incarcerated. As far as I could see, it was not for the benefit of the rank-and-file. We had used communication systems from the start, so if they had wanted to they could have asserted leadership years earlier.

When we had first arrived at the Hanoi Hilton, one heated debate erupted among senior officers over who outranked whom. Officially, the old-timers maintained the rank they had at the time they were shot down, which meant some newer shoot-downs outranked them because they got promoted before they got shot down. I could not have cared less who outranked whom.

That attitude was going to cost me.

In the fall of 1971, the four flights in our room each rotated our duties. When my flight was the duty flight for the day we were responsible for washing the dishes, sweeping the floor, cleaning the shitter, and rolling up punk paper for our cigarettes. About the punk paper: someone had devised a simple but genius way to turn our shit-paper into a cigarette lighter. We would roll the toilet paper into a skinny chain about thirty feet long and a quarter-inch in diameter, then light the end of it and blow it out, giving us a punk that would stay lit for hours.

One day, our flight was on duty and I was one of three guys assigned to wash dishes. The problem was that we only had a couple of hours outside in the afternoon and the cistern where we washed dishes was the same place where everyone washed clothes. A few guys were still finishing their clothes and we were just starting the dishes when the Vietnamese called out, "Go in!"

Al Stafford was right behind them, also insisting, "Go in, go in!"

We gave up on finishing the dishes. A couple of guys still washing their clothes also gave up and either flung the sopping clothes on the line or brought their dirty things in. But Dave Luna ignored Al and took his obstinate time about wringing out his clothes. It so happens this was the same Dave Luna who had marked time in The Rotten Cell long before my time, scratching six months of days into the bottom of a concrete bed. Apparently that ordeal had not been enough to squeeze the stubborn out of him. As the rest of us filed past Al to go in, we could see him giving Dave the evil eye for lingering. Al waited silently for about thirty seconds, and then his voice hit the air with a one-two punch: "Dave, go in!"

An hour later, Al called the entire room to a meeting. "Whenever I'm out there and I tell you to go in, don't give me static, don't ask questions, just go in. I have my reasons."

Dave Luna wasn't the only person who had problems finishing chores in the time allotted. I raised my hand, not to contradict Al, but to discuss the bind the dishwashing detail was in because we were last in line for the cistern. I wanted to ask if it might be possible to cut the dishwashers a break and allow us a few extra minutes outside. But Al ignored me and wrapped up the meeting. I didn't lower my hand.

He turned an exasperated look at me and snapped, "If you have questions, ask your flight leader!" He pivoted on his heel and marched into the head.

I clenched my jaw at being shut down in front of nearly fifty people simply for having a question that pertained to all of us. A few men gave me sidelong looks as the meeting broke up.

This was the last straw, proving to me the way senior officers handled junior officers' input: by minimizing our legitimacy. I had no recourse to address a superior officer about his rudeness. My face heated up as I realized Al had just shown me my place: the bottom.

I was pissed.

There was nothing left for me to do but cool off. So I did what I often did to blow off steam. There was a sixty-by-twenty-foot section in the center of the room with no beds and I walked the periphery. A lot of guys did the same for exercise or to prevent going stir-crazy. A few were walking around the perimeter at the same time I was, though most of the guys had moved to the beds to talk or to be alone with their thoughts. I made my first circuit and happened to pass by the shitter just as Al was coming out.

He walked toward me, apologetic eyes trying to meet mine. But I was still angry and his lousy timing just pissed me off more.

He didn't pay attention to my body signals: the long strides, lack of eye contact, red face. I knew that the message I was giving off would have been apparent to anyone: This guy needs space.

He drew close and repeated the same comment that had already ticked me off once, "Bob, if you have any questions in the future, bring it to your flight leader."

I gave him a dirty look and, not wanting to say something I might regret, stepped past him without replying.

Major Dan Sorenson happened to be standing nearby, and as I continued on I heard Al say to him, "Did you see that? He ignored me!" Then Al barked, "Wideman!" Several heads turned.

I turned my head too, but kept walking as I said, "Al, I'm in a bad mood now. Let's talk in five minutes."

On my second lap, he was waiting for me, back straight, arms folded, legs wide.

"I want to talk to you." His tone suggested, *You're in big trouble, young man.*

I continued to ignore him, walking past a second time.

By now the whole room was either staring or making a point of looking away.

"Wideman!" he shouted. "I want to talk to you."

"Al, let me cool down first."

He screamed at the top of his lungs, "Wideman, I order you to stand in the corner!" He pointed toward the closest corner to me, which happened to be right next to the latrine.

I stopped and stared at him, mouth gaping. I glanced at nearby faces, and they reflected my incredulity. Had a fellow countryman, a fellow prisoner, a fellow adult really just ordered me to *stand in the corner*? The North Vietnamese had often ordered me to stand or kneel in a corner, for hours or days at a time. No way was I going to let one of my comrades subject me to the same punishment! I was already a prisoner. Enough was enough.

I kept walking, not daring to look at him, afraid I might lose my shit.

I did cool off, but figured it wasn't wise to speak again until spoken to, so I kept strolling. I must have walked two miles around the room that afternoon.

Dan Sorenson fell into step with me. "Why don't we talk about this?" Dan was a quiet, patriotic, engineering type. Not the kind of guy I would look to for a heart-to-heart, but he was a decent person and a superior officer.

I shrugged. "Okay."

We walked to a far wall where there were fewer people. I sat down on one of the concrete abutments, and he sat on a nearby bed.

"Al would like to talk to you," he said.

"Look, I don't want to talk to the guy. He's out of line."

"I really think you've got to at least let him talk to you, you know?"

We went back and forth like that until I gave an inch. "Okay. You tell him that if he can talk to me in a civilized tone, I'll talk to him. And this talk won't be one-sided. If he insults me, I'm walking away. I'm not going to listen to it. Okay?"

"Okay." Dan was not the type for conflict.

After that, I lay down on my bunk and closed my eyes. Maybe when I woke, Al would have cooled off too. A couple of hours later, at about 8:30 in the evening, the guys were starting to put up their mosquito nets when Jack Rollins walked over to me.

"Stafford wants to talk to you."

I followed him to Stafford's bed, where Stafford did not rise to meet me but remained seated. Stafford didn't order me to stand at attention, so I stood at ease.

"Wideman." He spoke between tight lips. "I cannot and *will* not tolerate the level of insubordination you displayed today. I've already reported you to the wing commander."

I said nothing.

"That's all. Return to your bed."

I felt as if I had just attended my arraignment. I wasn't even supposed to *be* in the Navy anymore. My discharge date had come and gone long ago. I just wanted to keep my head down until I could get out of there. Our wing commander was Risner, who had more leadership experience than Stafford. I figured Risner would reprimand me in private and tell me to get my shit together.

I was dreaming.

That evening someone tapped on our wall with a message from Lieutenant Colonel Risner for everyone in the camp. A member of Stafford's information team read the message aloud to our room: "Lieutenant Junior Grade Wideman, you are a disgrace to the American fighting man. You have disobeyed orders, and you may face a court-martial when we return."

I could not believe my ears. I sat back hard on the nearest bed. I looked across the room at Al Stafford, who did not look triumphant. He looked scared, as if the situation had gotten out of his control. Maybe he feared he had taken it too far. I sure thought he had.

That was the only announcement. We were dismissed. Again I circled the room to walk off my outrage. This time Air Force Major John Pitchford from Natchez, Mississippi fell in step with me and muttered, "I don't think you're a disgrace. Just maintain your dignity." This meant a lot to me because John had the respect of all who knew him, so much so that his nickname was "The Sage."

"Thanks, Major. I hate to say it, but some of our leaders seem unbelievably screwed up." I walked on alone, making my way to

Ed Mechenbier's bed, where I said, "I'm thinking about sending a message back to Risner informing him that I'll file a complaint against him for defamation when I return home."

"I wish you would," Mechenbier said.

Galen Kramer was in a nearby bed and overheard us. "That would be great if you could, but no way is Stafford ever going to let you get on the wall with a message like that."

"You think I should just let it go?" My question was sincere.

"I didn't say that," Galen said. "This whole room thinks Stafford screwed you. A bunch of guys are saying the room is on the verge of mutiny because of what he and Risner did to you."

Galen's comment moved me. I had thought more people would fall in lock step with the party line.

I found out years later from someone on the covert information team that Risner had sent two messages that day. One was a secret message to Stafford telling him to get his act together and maintain control of his room. Risner had publicly called me a disgrace to the military, making sure everyone knew it, but privately chewed out Stafford for poor leadership, ensuring most of the prisoners would never hear it.

I was confident Risner had broken a basic principle of leadership by publicly excoriating me. A true leader praises in public and reprimands in private. From then on, I was skeptical of anything our leaders said. If the leaders in our camp could be so easily misguided, what other mistakes had our leadership made farther up the chain? I was not angry with Stafford any longer. He had let his temper get the best of him and he knew he'd screwed up. I didn't much care for him, but I figured he probably wasn't emotionally stable and I didn't expect him to be more than he was. But Risner? I had once admired him, and now he had done to me what the North Vietnamese could not: destroyed my faith in America's leadership.

21

COMMUNICATION

The U.S. Government occasionally found ways to sneak information to POWs via microfilm hidden in their packages from home. In the spring of 1971, word came down that one of the prisoners at the Hanoi Hilton had received a microfilm hidden in a plastic peanut. It contained information about us prisoners being a hot topic in international affairs. The way we heard it was that Neil Armstrong was traveling all over the world spreading the message that it was inhuman to hold U.S. prisoners in North Vietnam for so many years. For some prisoners, it reawakened hope, hearing that the first astronaut to walk on the moon, a guy who had started as an airplane pilot like us, was now publicly rooting for us.

Lieutenant Commander Doug Clower suggested we turn up the heat on the POW issue by enacting a moratorium on writing letters home. If our families found themselves in that sort of news blackout, they were bound to express concerns about our treatment, which many of us believed might put political pressure on the North Vietnamese negotiators at the Paris Peace Talks. Soon the official order for the moratorium came down through the comm teams. The plan was to do it in stages: *In the first month, 25 percent of prisoners will not write. In the second month, 50 percent will not write. In the third month, nobody will write. The purpose is to bring political pressure to bear on the DRV (Democratic Republic of Vietnam) in Paris.*

I was all for it. I would much rather give up writing letters for a year and be released soon than write a letter a month for the next twenty years and never see my wife or parents again. Our leaders made it clear that the moratorium was voluntary, yet by the third month all but a handful of prisoners had stopped writing letters. We all wanted to go home.

At first, the North Vietnamese acted upset over the moratorium. "We want you to write," the camp commander announced. Then, "We order you to write!" At one point we heard through our comm system that GTO went into one of The Heavies' rooms and "begged them to write."

The Heavies didn't budge. Why would they? The pressure seemed to be getting to the Vietnamese and that was the goal. The letter moratorium went on. I was good with that.

Then, about the middle of 1971, The Heavies changed the purpose of the moratorium. It was no longer primarily to put pressure on the DRV. They announced that the number one purpose was to demand that the North Vietnamese recognize them as the leaders of our 4th Allied POW Wing, and that the number two purpose was to get our captors to treat us according to the Geneva Convention. I didn't understand the need to throw in the extra demands. They were already treating us better than was required by the Geneva Convention, and so long as that was the case who cared whether they officially acknowledged our pecking order or not?

Instead of the Vietnamese wearing down, they simply tired of arguing with us. They took the attitude of *we don't care.* Between that and our leadership's new priorities, which weren't my priorities, I felt less incentive to stick with it.

After three months of playing along, I told Major Tom Sumter, Stafford's second-in-command, "I'm going to start writing home again. I don't believe in denying my wife letters just so Risner can use that as a bargaining tool in his negotiations with the enemy. I don't feel my treatment is bad enough to warrant that."

Tom gave me a song and dance about how the moratorium was working and the Vietnamese were only pretending not to care. "We need to stick together. I wish you'd reconsider, Wideman."

"Okay, I'll reconsider." After all, I had originally thought it was a good idea.

I spent another three months not writing letters, until one day the leadership announced out of nowhere, "We're going to start writing again." They made no announcement about whether or not we had met any of our objectives. The only thing the rank-and-file knew was that one day we were not writing, the next day we were, and that was that.

On October 16, 1971, the day before my twenty-eighth birthday, the guards told our room to roll up. I'd been waiting for this. For the past couple of weeks they had been trucking prisoners out of the camp, twenty to thirty at a time. All in all, about a hundred of us moved out.

I hoped we might return to Camp Faith, the place we thought of as the last stop before home. Instead, we returned to The Zoo. It seemed to me we were going the wrong direction, going back in time.

They put me back in Room 1 of The Office, the same building I had lived in before, but this time they had opened a doorway that had been plastered over the last time I was there, which turned my old room and the room next door into one big room. They put seven of us in there to start. Commander Al Stafford was the senior man. Russ Temperley was our second-ranking man. Then we had Joe Milligan, Jim Berger, Tim Sullivan, Tom Parrott, and me. The Office's four other rooms held six or seven people apiece.

Our corner room had a good view of a window from a nearby building called the Pool Hall, where the prison housed our Senior Ranking Officer, Major John Stavast. Thanks to the clear line of sight, our room became the communications link between our building and that building. Since Jim Berger and I were smallest, we were assigned to the comm team.

When it was my turn, I would sit on another prisoner's shoulders, poke my hands through the wide bars at the top of the windows, and use a form of sign language, with different hand gestures equaling different letters. The gestures had to be big so the receiving team could understand them from a distance. Stafford would roll his mat back and write messages on the floor with a brick tile he used like chalk. That way I didn't have to keep whole paragraphs in my head. Then I would relay the messages. This all required plenty of memorization and hand-eye coordination, as well as a certain amount of responsibility and risk. Taking on those challenges gave me a sense of pride.

I was not cleared for covert information, but Tim Sullivan, who did have clearance, told me I could pass those messages too because I wouldn't understand them anyway. He was right. The few times I passed along covert info, it was made up of letters, numbers, and phrases that made no sense to me—phrases like "Barkeep to stools." I had no clue what those messages meant.

Life was still relatively comfortable, almost too comfortable. We were settling down into a life that wasn't terrible, but it was still a life we didn't want. We still got out of our rooms twice a day for a total of five or six hours. In addition to chess, checkers, cards, and Ping-Pong, we now had a small soccer field, a volleyball and net, and a basketball and court complete with hoops and nets hung from what looked like two telephone poles—that was set up near the algae-coated swimming pool. We had a little vegetable garden in back that several of us worked on. Even Stafford tended to a little pepper plant. We continued to eat three meals a day. We still received packages and exchanged letters home.

The North Vietnamese started to bring a mobile library into the camp as well. I found reading a great way to make time go by more quickly. I read several books from famous Russian authors. I found Leo Tolstoy's *The Resurrection* fascinating.

Stafford took one look at that one and said, "Typical Russian book: gloomy, dreary, pessimistic."

I gave a chuckle of assent. "You're right. But it's well written, and I'm learning about the Russian Revolution."

As civil as Stafford and I tried to be, it was obvious we could barely tolerate each other. I found him increasingly annoying. He still couldn't sleep, which meant the rest of us couldn't sleep either. But I'd lived with nightmare roommates before. I figured I could survive this one.

Then, shortly after we arrived in the new camp, our door opened and The Boy Scout stepped into our room. Larry Bell had been our roommate for less than thirty minutes before he set to work getting on Al Stafford's good side. He didn't wait for Al to assign him duties before requesting them, making it clear he was ready to go above and beyond.

When he saw me hop on someone's shoulders to exchange messages with our neighbors, he looked so envious that I could see he wanted in on the action. I decided to buy a little good will and invited him to join the comm team, not thinking he might turn it to my disadvantage.

The first time Larry climbed on somebody's shoulders to exchange messages I could tell he was in heaven. For a moment, his grin made me feel like a good guy. He joined the rotation with Jim and me.

A couple of weeks later, right in front of me, Larry told Al, "You know, it'd probably be a good idea not to pass secret information through people who aren't cleared for it." Al agreed. Whenever secret info came in, Larry would whisper those messages to Tim. That cut Jim and me out of half the communications.

I let it go. I liked the job, but it wasn't worth a hassle.

***

Just before Christmas, word in the yard was that Joe Crecca, who lived in our building, had received a letter from his wife saying she had been in a car accident that put her in the hospital for a few days. She was home from the hospital, but still feeling blue, and Joe was worried. The next time the guys in our building went outside, he approached Stafford and asked him to send a message to Major Stavast requesting permission to record an audiotape to send home to his wife and daughter, something to reassure them he was going to be okay.

Stafford said, "I'm going to have to flag that message," meaning it would be kept secret, so that nobody but The Heavies and their covert communication team would hear about it.

Joe was overheard saying, "I don't care if people know."

"Yes, but some people might not understand why you're doing it."

So Stafford flagged the message, Larry sent it, and as far as the brass was concerned it remained hush-hush. It took a couple of days for Major Stavast to respond in code, and only a couple of minutes for the code to be translated into that day's gossip: "Sorry, according to our standing orders, I cannot permit you to make a tape home."

The way our leaders explained it, such orders were to prevent us from owing favors to the enemy or making propaganda for the enemy. But that hadn't happened in a long time. In this case, I thought it more likely Stavast was worried that if everyone asked to make a tape for Christmas the NVA wouldn't be able to accommodate them all, which would result in only their favorite prisoners making tapes. What's more, some of our leaders said even a tape intended to reassure our wives that we were okay could be propaganda, because it would make the Vietnamese look like they weren't bad guys after all.

I found Stavast's decision cold-hearted, but I didn't make waves. I liked Joe but figured he could take care of himself. Joe had quite the Italian tough-guy act going. One week my socks went missing, and I walked by his room to ask him to check his bedroll. His response was, "Get the fuck out!" Then he literally grabbed me and threw me out. I threw my sandals at his head, followed by a bowl, and we ended up wrestling on the ground until a bunch of prisoners ran over and pulled us apart. On the other hand, I also knew that Joe, who had taught me a lot about classical music, had a soft side, and it was clear he was worried about his wife. Tough or soft, he didn't cross The Heavies. He grumbled, but he obeyed.

A week later, a couple more guys asked Stafford to send a message to Stavast announcing their own intentions to make tapes: "We've been here now for five years and our wives haven't heard our voices. On the occasion of Christmas, we're making tapes home." In addition to actually wanting to send tapes home, it was clear they were showing solidarity with Joe.

Stavast's reply made his displeasure clear: "If prisoners send tapes home, it's against orders and it's going to undermine morale."

That's when our whole building protested, me included. We told Stafford to tell Stavast that after all Joe had done for his country he deserved to send a message to his wife.

Stavast finally relented and said Joe could make a tape, "But nobody else."

So Joe made a tape. And, despite Stavast limiting permission to Joe, so did two other prisoners. Then a third prisoner announced a plan to ask the Vietnamese to take a photo of him to send to his wife, who had not seen his face in five years.

Stavast was losing control of the situation.

When the three defiant prisoners didn't back down, Stavast used the comm system to send a public message in which he addressed the men by name and reprimanded them: "You have intentionally violated the orders of men appointed to serve above you, and what you have done is wrong. It could undermine morale. Because of your past excellent record and behavior, no action will be taken at this time, but any future acts of this sort will be dealt with."

That's about the time Howie Hill announced his decision: "I've been here a long time and I feel it's my moral responsibility to make a tape home to my wife." The leadership used the comm system to plead with him not to, but Howie decided enough was enough. He made a tape.

In response, Stavast sent two messages to Howie for all of us to hear. The first was more or less, "You are disobeying orders by accepting special favors from the Vietnamese." The second message came later: "From August 1st until September 1st, you are hereby suspended from all military duties." Most of us slapped Howie on the back and teased him about that one because it meant that for thirty days he would not be allowed to help us do chores, like washing dishes, cleaning the shitter, or sweeping. He got a thirty-day POW-vacation.

About four or five days into Howie's reprimand, he received a message from his friend Terry Boyer, who lived up the hill in another building: "I don't know what's going on down there, but hang tough. I don't hold anything against you for making a tape."

Some time after that, Howie sent a message back to Terry, and with that I learned something about Howie that I hadn't known: "Boyer, Happy Birthday and thank you for your kind words. After I made the tape, I got the first letter from my wife after not getting one for more than five years."

Five years! I could only guess what Howie had been going through. I was happy for him that the risk he took paid off with his wife. Others weren't so lucky. Soon word filtered through the prison that some of the guys' wives were divorcing them. Long-distance marital discord was spreading like an epidemic.

***

In the spring of 1972, American planes began to bomb North Vietnam again. At first, the raids were thirty miles or more from Hanoi. Soon they got louder and shook the ground more fiercely, and we knew they were getting closer.

One day the guys from our building were scattered around the courtyard, playing volleyball and basketball, gardening and bathing, walking and socializing, when an air raid siren screamed. A guard banged a club against a metal slab hanging from a tree limb. That usually served as our morning gong, but this time instead of the usual slow *bong...bong...bong*, he whapped it hard and fast: *pa-boing-pa-boing-pa-boing*. The guards herded us into our rooms. It wasn't easy: a handful of them moving more than two dozen of us, most of whom had grown complacent about air raids. The bombing was half over by the time they shoved us into our rooms and ran for cover.

After that happened a couple of times, the camp commander announced new rules: *Only two rooms allowed out at a time, one hour and a half in the morning, one hour in the afternoon.*

The guards started keeping our shutters closed and turning off our lights most nights. They dug a foxhole at the front of the camp, right next to the basketball court. Then they built a mount with a machine-gun placement facing the front gate. They had never done that before. Word spread across the camp that it could only mean one thing: They were not as concerned about an air assault as they were about Americans storming the gate.

By April of 1972, the American bombing raids were hitting nearby Hanoi. A couple of times it sounded like a locomotive was rolling right over our building. Explosions a mile away rumbled beneath us like earthquakes, windows sucked inward and threatened to shatter, plaster cracked, bits of ceiling fell around our heads. I felt plaster dust sifting into my hair and thought, If the raids don't work this time, our guys are gonna nuke Hanoi!

One night at about 2:00 a.m., the lights were still on when the air grew still as death. I heard what sounded like hundreds of feet tromping across the fields, then people screaming and shouting, then the intermittent storm of small-arms fire—*ta-ta-ta-ta-ta-ta-ta...ta-ta-ta-ta-ta-ta-ta*—and underneath it all, blood surging through my ears as my heart raced a thousand beats a minute, faster than gunfire. Good Lord, I thought, we're in the middle of a ground assault!

All I could do was lie there, wishing someone would turn the lights off, straining to hear the breathing of the men around me. Some still snored, oblivious. The screams and shots drew closer. Then one of the shutters near me creaked open. Here it comes, man. A guard is about to throw a grenade into the room, kill us all to prevent us from being rescued, because that's procedure. My heart stuck in my throat, breath stopped, gut clenched as I waited for the grenade to roll across the floor and explode, wondering if I had time to run, and if so, where? Nowhere. After surviving five long years of waiting, this was it. I was sure I was dead.

Fuck.

Then the shutter closed. Minutes later, silence fell. The shutter opened and closed again. That was it. We found out the next morning that it was a drill.

The next night they had another drill, but this time I slept right through it. Guess the human body can only stay on alert for so long.

About that time, the village next to The Zoo apparently had enough of the air raids. They boarded up their windows and left, the way people in South Florida would if a hurricane were coming. I had barely given the village a thought before, just distant men, women, and children, working, laughing, and playing. Now it was a ghost town.

We weren't far behind them. Not long after the villagers vanished, the guards moved us all back to the Hanoi Hilton. Once again, it seemed like we were moving in the wrong direction. Hanoi was a primary target for bombing raids, which swept across the city almost daily.

One morning, after an air raid the night before, Stafford asked me to relay the following message to our neighboring building: that during the raid a guard had climbed a ladder up to our window, poked a machine gun through the bars, and aimed it down into the room where we lay in our beds. What had in fact happened was that a guard did climb a ladder up to the window and peek in, and then when the air raid arrived he got down. He had a rifle slung over his shoulder, but he never aimed it into the room.

After I delivered that false message and climbed down from the shoulders of the prisoner who was holding me up to the window, I asked Stafford if I could talk to him in private. We stepped to a corner, our eyes darting to the other men standing a few feet away. It wasn't very private.

"Al," I said. "It's very difficult for me to pass on information that isn't true. If it's okay with you, I'd just as soon be dropped from the comm team."

He didn't look surprised at all, just said, "Talk to Larry Bell about it."

I crossed the room to The Boy Scout. "Larry, I don't want to be on the comm team anymore."

"Okay."

That was that. I was out.

I was sorry to give up my spot on the team. Sharing in that extra responsibility and risk had made me feel that even though I was a prisoner I still had something to contribute. But I wasn't proud of passing on false information, even false information about the enemy.

About a month after the move back to the Hanoi Hilton, half or more of the camp's prisoners climbed into covered flatbeds again for another mystery ride. We weren't the only ones hightailing it out of town. The guards told us that Vietnamese citizens were evacuating from Hanoi by the thousands, for fear the U.S. military was going to level the city.

The guards told us, "Hanoi's not important. We expected Hanoi to be gone long ago. America will still lose." The point was they would never give up.

I believed them.

22

DOGPATCH

This evacuation was nothing like the others. The slow journey out of the besieged city was only the beginning. For most of the trip, the convoy of trucks strained to drag us up narrow mountain roads with steep drop-offs. It took some twenty-four hours, from early morning dark to early morning dark. We arrived at about 2:00 a.m. at a small, new camp high in the mountains, within twelve miles of the Chinese border.

A guard with a kerosene lamp led eighteen of us into a grim building with walls of foot-thick stone. At first, the building appeared to be divided into two rooms with just enough room for the eight plank-beds forced into them, mere inches of space between. Each room was lit by a single dim lantern, but despite the poor lighting I could still see how cramped and dirty it was. The thought of living in such a packed room filled me with claustrophobia. Maybe if I stayed in there long enough I would snap.

Then I found two smaller rooms between the two main dorms, one-man cells with a little more breathing room. I put down my bedroll in one of them, feeling lucky to snatch some space to myself. My luck only lasted a few minutes.

I was standing in the room I had staked out when Al Stafford and Dan Sorenson walked in, talking about our new digs like a couple of homebuyers at a low-budget open house.

Sorenson looked around my room, nodding appreciatively, and said to Stafford, "Hey, this is a pretty nice room," right in front of me, like I wasn't there.

"You should have it," Stafford said, also ignoring me even though I was only standing a foot or two away. It felt as if they were daring me to react.

"Would that be okay?" Sorenson said.

"I don't see why not. You're second in command. You handle it. As far as I'm concerned, you can do anything you want—except I want that bed right there." Stafford pointed at a bed in one of the eight-man rooms. Of course. How else would he have somebody to talk to at night?

Their little exchange blew my mind. They could see I already had my gear in that room.

Sure enough, the next day Sorenson ordered a room switch. He moved me into the smaller solitary room and put himself in the bigger one. As far as I was concerned, he got that room simply by pulling rank, which he had the authority to do, but the way he did it really ticked me off. What a way to lead.

The room switch was hardly the worst news I heard that day. The guards told us we would not go back to Hanoi until the war was over. With that news, my mind started chasing its own tail, my worst fears circling again and again: Maybe they put us up here so they could do whatever they wanted to us and nobody could ever find us. Maybe they expect the war to last another twenty years. Maybe this no-man's land is our punishment for the bombing of Hanoi. Maybe they'll never let us go home.

The promise of a final trip to Hanoi at the end of the war became nothing but a dream I didn't trust. Hope became painful. In that frame of mind, having either of the private rooms to myself might not have been the best thing for me after all. I shut my door and wouldn't come out.

Ed Mechenbier came by and spoke through the door, "Who's in there?"

"It's me."

"Oh. How you doing, Bobby?"

"I'm okay." I choked on the phrase, damn near on the verge of tears because I was back to thinking of myself like the Count of Monte Cristo, forgotten in his dungeon.

I didn't come out until the next day. After that, I held onto my sanity by just focusing on each day in front of me.

We had a dim and narrow courtyard outside, with a couple of slimy cisterns containing suspect water. The guards brought us wire and bamboo poles so we could create clotheslines, then laid bamboo slats over the top of it all and covered them with brush to camouflage us from aircraft. In those first few days, the guards sent a few of us uphill with buckets to fetch water, one at a time. I carried my bucket to a small clearing where a mountain stream paused at a clean pool. I stepped to the edge of the steep slope to look out from our perch at the lush fairytale mountains surrounding us, the terraced farm below, and the valley opening up beyond. Except for the ten concrete prison buildings, I might have thought I was in Paradise.

We called it Dogpatch.

With that view I began to relax.

The guards seemed to be doing their best to ease our shock and uncertainty. They chopped down yet more bamboo to make troughs to carry the water from the uphill stream down to the cisterns, so we wouldn't have to keep hauling buckets—though I sure wouldn't have minded hiking up to see the view. Every day the guards opened the doors of the building to give us access to the courtyard for most of the day. The food improved greatly: once a week we had buffalo meat with gravy, which was delicious, and they started mixing maize into our rice.

We still received letters and packages from home. Pat was getting better at those, filling them with hard candies, puddings, nuts, little cans of Vienna sausages, soap, toothpaste, photos, and Flintstones vitamins. The guys still played chess, checkers, and cards, though the courtyard was too small for the active sports we had learned to enjoy.

I played endless hands of gin rummy with Dick Tangeman, who had been a radar attack navigator for a Navy A-5 Vigilante. All four of Dick's wisdom teeth were impacted, cracking under pressure, no dentist around to help. He wrapped his head in a damp towel to ease the pain, like a kid with mumps. During our card games, he paused every few minutes to say, "Excuse me," and then spit pieces of broken teeth into his hand. I never once heard Dick complain.

After my first couple of nights in the new camp, we all decided it was only fair to turn my small room into a rotating accommodation for everyone but Stafford and Sorenson—who liked where they were. That way each prisoner would have one night of privacy every sixteen days. It wasn't quite enough relief for me, especially since Stafford kept us up most nights with his nonstop chatter.

The Boy Scout was in the room next door, but it seemed I could never completely get away from him. One day he was communicating with another building, using hand signals, and Jim Shively was sitting nearby. Larry didn't know that Jim could read his hand signals. Jim later whispered to me that Larry had signaled, "It's good to be talking to patriots." Jim said it looked clear to him that Larry was comparing the guys in the other room to the guys in ours. How insulting! It sounded hypocritical too. When The Boy

Scout had first moved in with us back at The Zoo, I had asked him what he thought about the prisoners in our building, and he had said, "I have to admit, these guys are more intelligent than the average prisoner." Apparently educated men who dared think for themselves were not patriots in his eyes.

I did feel sorry for Larry the morning he woke up with his scrotum swollen to the size of a grapefruit. He had accidentally rolled over onto one of his testicles in his sleep. He was obviously in a lot of pain, and he spoke very little over the next few days while he waited for the swelling to go down. Six years of sleeping on a board night after night was bound to do a number on one body part or another, and Larry Bell was no exception. Most of the time though, between Bell and Stafford, I was going crazy trying to contain my temper. The other prisoners could tell I was about to lose it, so they gave up their rotation nights in the single room and let me live there by myself. I owe my sanity to their generosity.

One evening the guards let about a hundred of us watch a movie in a big assembly hall they had built from bamboo. The movie was a full-length, black-and-white Vietnamese flick about the war, but in this film the Americans were the bad guys and the Vietnamese were the heroes, complete with their very own John Wayne. I'll never forget one of the scenes where the villagers hid in the bushes and then the Americans found them. Next, the Americans dragged a screaming old Vietnamese lady away through a swamp.

Another night, the guards told us to make sure we put up our mosquito nets before we went to bed because there was a typhoid epidemic in the next valley. Chief Warrant Officer John William Frederick Jr. contracted the disease. The North Vietnamese immediately put the Marine in a truck and drove him to a hospital in Hanoi. It was no use. He died before they could get there. At the time, I thought of him as one of the old-timers. He was not quite 42.

That fall, another American presidential election came and went, my second as a POW. Richard Nixon won by a landslide. During the campaign season, we received an order from the senior ranking officer, Stavast. He told us that our order of loyalty was as follows: unit first, country second, family last. That was not the order of my priorities. I believed family came first, then country, and finally my unit. Just one more way I thought our leadership's priorities were upside down.

Around that time, we had a camp shake-up. The guards moved our group into another building. This one had two rows of two-man rooms. A few prisoners I hadn't yet met moved in with us. One was Air Force Major Bill Baugh, our building commander, a laid-back officer who made it easy to forget rank. He was generous with everyone.

The guys in my last room had always shared packages from home. The new guys lived by the unspoken policy that each prisoner kept his own package, which was kind of sad for guys who didn't get much. Bill was the exception.

Bill's wife always sent coffee in his packages. The Vietnamese let us have a small makeshift stove made from a tin can, with a grill on top and a section cut out at the bottom to burn wood. Bill made coffee on that little stove, and just as the roasted smell filled the room, he always called out without fail, "If you guys want some coffee, come on down." Shared his sugar, too.

Bill also shared information that ended my ignorance about a mystery that had long been eating at me. He had been one of the prisoners in the room at The Zoo where John Dramesi and Ed Atterberry attempted to escape in 1969. I had always been curious about that day, which I had only pieced together from the odd glimpses I caught from solitary, the bits and pieces George and

Larry told me, and the scattered messages and rumors that came through the walls. Bill filled in a little more, but he seemed to find it hard to explain in a way that would make me understand the feelings underlying the facts.

"You'd have to know the personalities involved to understand what happened," he said.

He explained that for John Dramesi there was no question about it: he just had to escape. Of eight prisoners in their room, six were adamant they did not want to go with him because of the odds against success. As for Atterberry: "Dramesi just talked him into it," Bill said.

He said Atterberry died in that escape attempt. I never did find out whether he was killed in the jeep crash I thought I heard that day. I was tempted to blame Dramesi, to say Atterberry's blood was on his hands, but Bill suggested that each man had only done what was in his nature. It seemed Dramesi was the sort of guy who would chew off his own leg to escape a trap, while Atterberry was the sort of guy who was easily persuaded in the heat of the moment. The other six were realists. Among those six, Bill weighed freedom against survival, and survival won the argument. The sobering thought was this: Atterberry was no longer in prison, but he was also never going home.

I liked Bill. He accepted people as they were.

Not long after the U.S. presidential elections, one of the guards took me in for quiz, this time with Squirrel. Instead of interrogating me, he handed me a letter and told me I should read it right there. I knew they read all our letters before we did, so I barely gave that a second thought. Still, it was unusual for the Vietnamese to expect us to read our news from home right in front of them. What could be the problem?

The letter was in Pat's handwriting. I relaxed. Then I began to read and it slowly dawned on me that I was reading a Dear John letter from my wife. It started happy enough. She wrote that she understood the war would be over soon and that she was glad I would be coming home. But she added that we needed to "talk about things" when I returned.

She recapped our six years of correspondence from a point of view I had never considered. She said that after I was shot down, she waited desperately for an answer to every letter she sent, but that my letters were few and far between. She said that in each letter my words felt increasingly cold and distant. I had thought that, by keeping my letters simple and not saying much, I was protecting her from excessive bad news or from too much hope. Her interpretation of my brevity was that I didn't love her anymore. What's more, she said she was no longer sure about her own feelings. She said we knew so little about each other's lives anymore.

She ended her letter with words to the effect of: "In view of the circumstances, I don't know if it would be wise to continue this marriage after you return."

What I didn't realize then was that Pat had consulted a divorce attorney and he was responsible for much of the language in the letter. I didn't know he had charged her three hundred dollars for this "service." In any case, it was clear she was preparing to divorce me.

This was uncharted territory, but I knew one thing: I did not want my marriage to end. I needed to hold on to the idea of Pat waiting for me. The hope of returning to her was the only thing I lived for, the only thing that made the interminable waiting bearable. She had become as indistinct as a dream, but it was the only dream I had. My guts felt like they'd been scooped out.

I looked up from the letter to see Squirrel studying me with concern. "What does it mean?" he asked.

"If I'm released today, it will be okay, but if I'm here for four or five more years, it's probably not good. My wife will leave me." I didn't want to admit the possibility that it was too late already.

"Do you want to write a letter home?"

I didn't hesitate. "Yes."

"I'll talk to the commander."

He sent me back to my room, where I waited for the camp commander's decision.

Christmas was coming, and soon the camp commander announced that, in honor of the holiday, anybody who wanted to make a tape to send home had permission to do so. I figured nobody could accuse me of accepting special favors from the Vietnamese since they had made the offer to everybody in the prison. So the next morning, when the guard opened our door to let us out to empty the shit cans, I stepped outside and said to him, "Wideman. Tape."

The guard nodded and left. He returned about two hours later and gave me a couple of long sheets of paper. "First write your letter here." This was standard procedure. Once my letter was finished I would read it into a tape recorder.

My roommate, Ed Mechenbier, spent the day with me, helping me compose a letter to convince Pat not to call off our marriage. It was strictly a letter to my wife. I wrote nothing about the war, about my treatment in prison, or anything that could be construed as propaganda.

That afternoon I was sitting alone in the room poring over my words when Bill Baugh walked in and sat on the bunk across from mine. "Somebody said you asked to make a tape."

My whole body tensed. "Yes, I did." He didn't have to tell me who said it. I felt pretty certain it was Larry, because he was the only one I had seen nearby when I spoke to the guard.

"Why didn't you tell me?" Bill said.

"First of all, I didn't want to get you involved in it. Second, I'm not begging Stavast to let me make a tape to my own wife. There's nothing political in this." I waved the letter at him.

He pursed his lips as if considering my words, but said nothing.

"I think the policy is wrong," I said.

"Yeah, you've got me there. But you know, you should've told me about it."

"Sorry about that," I said. Bill was a fair man, and I felt bad for not trusting him to support me.

He nodded. "If you want to write a letter or make a tape, that's fine with me. But any time you want to do something like this in the future, just tell me. I don't want to hear it from somebody else. I'd like to hear it from you."

"Yes, sir."

He asked me to send a message to Stavast saying I had personal problems at home and to request to make the tape. Bill added his own message: "It certainly seems that his reasons for wanting to make a tape are legitimate and I recommend he be allowed to make a tape home."

I didn't wait for Stavast to reply. The next day a guard took me to a room where Squirrel and a microphone sat waiting for me to deliver the letter. Squirrel started the tape, the wheels began to turn, and I began to speak. "My darling wife Pat, I have received your letter, and I just want you to know I love you..."

When I returned to my building, Air Force Major Glen Nix asked to talk to me. He was second in command. We had a good conversation. He said The Boy Scout told him I made a tape. I told him I did not say anything on the tape that would harm the United States. He let it go.

As for whether I had said anything on the tape that would save my marriage, I would have to wait. It might take weeks, even months, to receive Pat's answer. For more than five years, the only thing that had kept me going was the ever-dimming possibility of going home. But going home to what? I wasn't sure anymore.

23

A HERO FOR WHAT?

On December 13, 1972, peace talks between the U.S. and North Vietnam broke down. President Nixon ordered a massive bombing campaign to break the stalemate. For two weeks during the Christmas season, American air assaults dropped more than twenty thousand tons of bombs on the cities of Hanoi and Haiphong. We were too far away to witness any of it. Up at Dogpatch, we just waited.

I still lived in a two-man room with Ed Mechenbier. I rubbed his back every night to help ease the pain from his abscessed tooth, as his other roommates had before me.

Early one night in January of 1973, we heard trucks roll into the camp. I stood on Ed's shoulders so I could see through one of the vents above our door. I saw a couple of trucks on the hill right below us. Another truck drove up and parked behind them. Then another, and another.

"I count fourteen trucks!" I didn't bother to lower my voice.

"We're going home!" Ed said.

I jumped down and we gave each other a bear hug.

It was harder to get our arms around the hope, after so many false starts. Going home. Sometimes home barely seemed like a real place anymore. Other times it seemed like the only thing that was real, and our lives in prison the illusion. Guards strode through our building, telling us to pack our belongings to return to Hanoi. They had said we would not return to Hanoi until the war was over. So, yes, we must be going home.

Right?

A few hours later they loaded us into canvas-covered trucks and we headed *down* the same mountain roads we had driven *up* several months before. Bill Baugh and I were in a truck together, along with twelve other prisoners. We had only been on the road for about an hour when I started feeling nauseous from the diesel exhaust, which flowed right up into the truck bed and got trapped under the canvas. Soon Bill was complaining of nausea too. The guards refused to remove the canvas, and just gave us a bucket to throw up in.

After an hour of puking with no relief in sight, I'd had enough. I stuck my head out from under the canvas at the back of the truck to breathe some fresh air. The guards went nuts, shouting, pounding on the truck bed, smacking my legs, and pointing rifles at me. Then one of them saw me puke out the back. This seemed to rouse his sympathy. He spoke to the other guards and they left me alone for a couple of minutes. After that brief respite, one of them ordered me back under the canvas. Bill and I continued to vomit into the bucket all the way back to Hanoi. The other prisoners in our truck did not need to use the bucket.

Although it had taken a full day to get to Dogpatch, it only took four hours to return to Hanoi, downhill all the way. We reached the surreal outskirts of the bombed-out city before sun-up. We passed the jagged silhouette of the wrecked Doumer Bridge, like the cracked bones of an ancient dinosaur strutting up into the sky, decaying in the dark before dawn. Workers on cranes were busy welding the bridge back together, and the bright flashes of welding arcs amid dripping metal lit the night like falling stars. The workers were listening to music over a loudspeaker, and an eerie song featuring a one-stringed instrument called a *Dan Bau* drifted to us over the grinding of the truck engine.

We arrived at the Hanoi Hilton to the sight of dawn breaking on the horizon and a rousing welcome from our fellow prisoners who had remained behind. Everett Alvarez was hanging out his window shouting my name, and my face stretched into a grin broader than it had in years. He wasn't alone; dozens of prisoners were hanging out of their windows cheering and shouting the names of friends.

Returning to the big rooms was like entering a high school cafeteria as the few men who had been left behind during our stint in the mountains caught us up on the terrifying, exhilarating action we had missed while we were gone. American B-52 bombers had pulverized Hanoi. North Vietnam later said that more than sixteen hundred civilians were killed in the bombing of Hanoi and Haiphong. The U.S. said it lost fifteen B-52s and eleven other aircraft. The prisoners said that George McSwain had hung out his window during the air raids, hooting and hollering encouragement into the sky. Like a madman.

Not long after our celebrated return, the party was over for me. Bill Baugh handed me a written message from Major Stavast. In the note, Stavast told me that he was responsible for the actions of all the prisoners under his command, and reprimanded me for making a tape to send home. I felt no shame, only disappointment at another military officer who seemed more concerned about his career than anything else. Bill tried to soften the blow, telling me it was just a by-the-book interpretation of my actions and that I shouldn't take it too hard. He said it wasn't the sort of thing that was going to ruin the record of a POW, not after I had sacrificed so much.

It was hard to dwell on disappointment for long. The news was real: the war was over. North Vietnamese officers walked from room to room and read the Paris Peace Accords to us. They told us they would release us in groups, one transport a month in exchange for American troops withdrawing from South Vietnam. We would be released in order of our shoot-down dates. The wounded would go with the first group.

The first wave of 115 prisoners went home at the beginning of February 1973. I was scheduled to go home with the second group of 115 in March.

The Vietnamese gave back some of our belongings that they had confiscated when we were captured. Married prisoners got their wedding rings back. I had never worn my ring when I flew. It must have been sent back home years ahead of me.

While we waited to be released, the Vietnamese let us outside almost all day long. We mingled with prisoners from other rooms, and for the first time I talked with several prisoners with whom I had communicated through the walls over the past six years but whom I had never met in person. They were like old friends with new faces.

For the first time in six years, I had an opportunity to talk freely with Jim Pirie, the friend from my squadron whose feet had been saved by that goofy pair of boondocks he insisted on wearing. Now we were both wearing rubber sandals—his boondocks and my high-topped boots long gone. We talked about all the people we had met, and how sometimes it was harder to cope with our comrades than with the enemy.

"You know," I said. "I can probably count on one hand the guys I want to see again after we're released."

"Yeah," Jim said. "A lot of these guys are real assholes, aren't they?"

One day, a couple of Vietnamese officers came to our room and announced that a special group of prisoners was going to be released before the larger March group, as a show of goodwill to the United States. One of them then read the list of twenty who would be accorded this privilege. The list included a few of the guys who had become my teachers during my time in prison, teaching me about history, philosophy, political science, music, and the arts. I was surprised to hear my name among theirs.

Before I had time to absorb that they were sending me home, Larry jumped to his feet, indignant. "Hey, those are the prisoners from *that group*!" He was referring to the men he had once told me were smarter than the average prisoners, the ones he had implied were not true patriots. The instant Larry said that, the officer announced Larry Bell's name as one of the special twenty. Larry's mouth slammed shut as he looked around the room, apparently embarrassed. It was hard not to laugh at him, this Boy Scout who was so sure he could distinguish the deserving patriots from the undeserving faithless. He did not say much on the subject of patriotism after that.

I soon put Larry out of my head. I was overwhelmed by the idea of seeing my mother, father, and brother, and hoping that when I saw my wife she would remember she loved me. I would tell her again that the hope of coming home to her was what had gotten me through the long days and longer nights that I had spent without a home, purpose, or future. I prayed, in the unspoken but repetitive way of a man unused to praying, that I still had a future.

The guards moved the twenty of us who were going home early into a room together. Spirits were high, everyone talking about home. It seemed to me we tried to keep from getting emotional by talking less about our wives and mothers, and more about things like TV, air conditioning, hamburgers, cars, women in miniskirts, and American music—the Beatles and Elvis, The Rolling Stones and Frank Sinatra.

The night before my early-bird group of twenty was scheduled to go home, a couple of Vietnamese officers came to our room with another announcement. This time they said that one of us would have to give up his seat on the next plane home and be replaced by Jim Bailey because Bailey's father was dying. "Bailey in, Wideman out," one of the officers said. They led me out of the room before the news could quite sink in. I had been in, and now I was out, and in both cases I had no idea why.

Jim Shively stepped up to me as I was leaving and said, "You're the lucky one."

I had no idea what he meant. I never did find out.

The Vietnamese put me in a different large room with a new group of fifty guys, where Colonel Norman Gaddis was the senior officer. I liked Gaddis. He seemed like a good leader, not constantly pushing some agenda.

The next day, a North Vietnamese photographer came to our room and asked to take a group picture of all fifty of us. Gaddis would have nothing to do with that. First, he insisted we not go outside. Then, when he could not prevent the photographer from coming in, he bellowed, "Soldiers, turn your backs to that camera!"

We had no choice but to obey.

The Vietnamese officer in charge turned a stunned face to Gaddis. "Why are you doing this to me?"

Gaddis ignored him.

I didn't see the big deal in letting the Vietnamese take our pictures now that the war was over. I'll never forget the painful rope tricks the NVA pulled on us that first year, and I thought I was going to lose it being trapped in small rooms with two or three fellow prisoners for the first three years, but for the final three years the treatment had ranged from good to low-budget country club. I was ready to move on, and I didn't understand why our senior officers were still intent on making the enemy look bad.

I had heard that a few senior leaders had been punished more severely than the rest of us in the early days, so maybe they had a legitimate beef. On the other hand, I can also imagine their egos played a role. Some tried to feed the rest of us stories that they were mistreated because they had more valuable intelligence than the rest of us. By then I knew that wasn't true. No warrior shot down on the front line had the kind of information interrogators wanted.

Shortly after that, the guards called me out for an interrogation. The interrogator was a high-ranking officer with a tailored uniform. He said Gaddis was out of line when he prevented the Vietnamese from taking pictures. "I think Colonel Gaddis overreacted."

"Maybe," I said, neither agreeing nor disagreeing. Although it did seem to me that Gaddis overreacted, I didn't know his personal history and didn't think it was my place to judge either way.

It turned out that wasn't the main reason this officer wanted to talk to me. He asked, "Do you know why you were the one bumped from the group that is going home today?"

"No, I don't."

"It was not our choice to bump you. Your government told us to bump you."

"Really?"

"Do you know why?"

"No. Do you?"

"No."

We studied each other. I wondered if he was telling the truth. Maybe he wondered the same about me.

He sent me back to my room, where I relayed to Gaddis everything that had transpired in the interrogation. Gaddis thanked me for the information, but added nothing to enlighten me.

I wasn't the only one thrown off by the politics of who goes home first. The pilots that arrived to fly the group of twenty back to the U.S. had to come into the camp because the twenty prisoners refused to leave. The group said they did not think it was fair for some of them to leave ahead of other guys who had been shot down earlier. We were all in solidarity on that one.

The pilots of the waiting plane insisted the men had to relent. One of them explained, "We understand, but look, you have to get on this plane. If you insist on staying here after the Vietnamese offered to send you home early, we're looking at an international incident. Things between the U.S. and the Vietnamese are delicate right now."

Jim Pirie was the ranking officer among the twenty. Pirie, who would have made a good diplomat, sent a message to Gaddis that the twenty men would not get on the plane home unless Gaddis *ordered* them to do so. Gaddis got the picture. He ordered them onto the plane. The twenty obeyed Gaddis's order and left camp with the pilots.

I stayed behind.

After the twenty departed, the Vietnamese let us outside to resume our usual sports, games, and conversations. I wandered around a bit, not knowing what to do with myself. I found myself standing near someone I had not seen since my capture, Navy Commander Jim Mell, who was my executive officer at the time I was shot down. I turned to him and smiled. He only nodded politely as if he didn't know me, and turned away to watch some game or other.

So I said, "Hello, XO."

He turned to me again, squinting. "Bobby? I didn't know it was you. I can hardly see you. My eyes have gotten worse."

"It's good to see you, sir."

That's about the time Loren Torkelson walked over to us and, to my surprise, addressed me instead of the commander. "What in the world did you do?" I couldn't tell if he was accusing me, praising me, or just plain curious.

"What do you mean?"

"You're a hero."

"A hero for what?"

He smiled and shook his head. "You're just a hero."

Loren never told me why he said that. Was I a hero because of what Risner and Stafford had done to me? Was I a hero because he thought I was the one who had bounced myself from the group of twenty that went home out of shoot-down order? Was I a hero because of something else I had done during my six years as a prisoner, something that nobody else had done?

I would never know the answer. After that, though it would always make me uncomfortable for someone back home to call me a hero, it would make me smile to think of Loren saying it, a guy I barely knew, but a fellow POW just the same.

24

FREEDOM

The day my group of 115 POWs was scheduled to ship home, the guards took us to a separate part of the camp, where we stripped to our underwear and waited to be fitted with dress shirts, slacks, and shoes. The black dress shoes the guards gave me were too tight, which felt all the more uncomfortable after years spent in sandals or bare feet. But I figured I could slip them off on the plane, and after that who cared? I was going home, where I could buy new shoes. They also gave us a duffle with toiletries, cigarettes, and room for the few small possessions we had collected during our time in prison. The only private possessions I packed were photos and letters from my wife and family; I was not eager for souvenirs of my time in Vietnam.

The guards walked us to several buses parked on the street. Hundreds of Hanoi's civilians lined the street as if waiting for a parade. Some waved and smiled, others stared hard as if hoping to set fire to us with their eyes. But none of the city-dwellers threw things or beat us. It was a far cry from the reception the villagers had given me when I fell from the sky into their world. We filed onto the buses, about twenty men per bus. The Vietnamese gave each prisoner a whole seat to himself, even though each seat could have held two or three people. I don't know if they were trying to give us first-class space to make up for years of crowding, or if they were just worried that packing us in might risk the boys-will-be-boys behavior that could lead to a riot.

We rolled out for the forty-five-minute drive to Gia Lam airport outside Hanoi.

At the airport we sat on rows of benches in a large waiting room, while American music played over a loudspeaker and smiling Vietnamese men walked up and down the aisles handing out bottled beer and shortbread cookies. A half hour later, they loaded us back onto the buses and we rode another ten minutes to the prisoner exchange point, a large tarmac where a troop-transport plane waited like an enormous white whale. There must have been a thousand Vietnamese and Americans milling about—military, prisoners, civilians. The buses stopped, and we got off and waited in columns of twos for our names to be called over a loudspeaker.

When I heard my name, I was not thinking, "I'm free!" but only, "Please don't let me trip and fall and make a fool of myself with all these people watching."

I walked toward a desk, behind which sat a North Vietnamese officer I recognized. He shook my hand, the first time a North Vietnamese had done that since the smiling guy who shared his vitamins with me six years before. The brief grip of the officer's hand was neither friendly nor unfriendly, just a strange point of contact between two people whose relationship had changed overnight because of decisions made by people half a world away. He handed me over to an American officer nearby who escorted me fifty yards to the waiting Air Force C-141. We walked around the rear of the plane and up a gangplank into its belly, where Air Force nurses stood waiting. Actual American women. One of them escorted me to my seat. I floated down the aisle as if in a dream.

The engines roared to life, and I held my breath as the plane rolled down the runway for takeoff. It wasn't that my crash six years earlier had made me afraid to fly, but that over the past six years I had become fearful I might never escape this place. The plane lifted smoothly into the air. Still I barely breathed. I was waiting for one more thing. Then I heard it: the landing gear lifted and locked into place, the signal that there was no turning back now. I was free! Every taut muscle in my face released. A nurse who sat facing me grinned. "What a look!" I had never cried in front of another human being except my mother and the men who gave me the rope-treatment after my capture—an automatic response to pain. But at that moment I could not hold back tears of joy.

The date was March 4, 1973. I was 29 years old.

The flight to the Philippines was peaceful. I looked down at the expanse of the sparkling Pacific, and it seemed like a different ocean than the one that had passed beneath me on that final bombing run back in 1967. I felt grateful for my freedom, as if experiencing it for the first time.

At Clark Air Force Base, we were instructed to pause one at a time at the airplane door until our names were called and then step down the stairs. At the bottom, an admiral wore a stiff smile, shook our hands, and motioned us to a waiting bus. An American flag waved from a pole next to the bus door. One by one, the men in front of me saluted the flag. I saluted when it was my turn, feeling a strange combination of homesickness for my country, anger over the years I'd lost far from America, and embarrassment over feeling that we were all on display.

The bus took us to the base hospital, where I saw several injured prisoners who had been captured in South Vietnam and Laos. They were in terrible shape: bent, emaciated, scarred, eyes staring inward. Several lay on gurneys. A few were hooked up to feeding tubes. Viet Cong guerillas had kept them in dirty make-shift camps or cages for years. The treatment of those of us shot down in North Vietnam was a picnic compared to the treatment of those caught in South Vietnam or Laos.

In 2015, officials at the Robert E. Mitchell Center for Prisoner of War Studies in Pensacola, Florida told me that 591 prisoners came home when the Vietnam War ended in 1973. Forty-four prisoners captured in South Vietnam or Laos died in captivity while twenty-eight prisoners captured in North Vietnam died in captivity. But of the pilots shot down in North Vietnam who made it to Hanoi, only seven died in captivity. Long before I knew those numbers, I felt my luck at having wound up in Hanoi, but at the same time I still felt my lost years pressing on me.

Thanks to my tight shoes, I couldn't help but limp around the base hospital, though I tried not to because I didn't want people to assume it was because of a war injury, especially not in this facility full of men who had been through so much worse than I. Nobody asked about my limp, so I didn't get the opportunity to explain.

The guys from our plane went to the cafeteria and ate a huge meal, surrounded by news photographers snapping pictures, bulbs flashing like an indoor lightning storm. Once I started eating, I barely noticed them. Every American dish we could imagine was available, but most prisoners wanted breakfast. I ate steak and eggs. I did not see a soul eating rice. It wasn't gourmet cooking, but I've never enjoyed a meal more.

After I ate, I talked to every person I could, trying to learn what had been going on for the past six years. I was starved for news: big news, small news, any news. One fellow prisoner told me that The Boy Scout's wife had divorced him. My good mood faltered. "I'm sorry to hear that," I said, and I really was. I not only felt bad for The Boy Scout, but also for myself, knowing I might be next. Major Kasler told me that all was good with his family. I was happy for him and grateful for the hope this gave me.

I still didn't know what my marital status was: dead or alive.

I phoned my wife. She reiterated the theme of her letter, reminding me in a hesitant voice, "We have some things we need to talk about when you get home." I suggested we save it until then, hoping that when we saw each other the years would melt away and the old feelings would return.

My dad phoned me, and I said, "I'm fine," unable to get many words out beyond that. The stern figure of my youth sounded so shaky on the phone it was hard to believe it was him.

Rest and relaxation wasn't the only thing on our agenda at Clark. The Navy quickly fit me for a uniform and gave me a military style haircut for the journey home. After that, I visited a doctor. I weighed in at 125 pounds, just ten pounds less than when the North Vietnamese captured me. I saw myself in a mirror for the first time in a long time and thought 125 pounds looked just about right on me.

The biggest complaint I had for the doctor was that my ears had been ringing for three years. I hadn't thought much about that persistent hum until that moment. It had slipped into the background of my life. The doc gave me a long look and had me lie down on a table. He balanced my head over a metal tray, picked up a syringe of warm water, and flushed both ears. I heard a loud *pop!* and felt the sudden suction and rush of a vacuum releasing. He showed me the tray. Each ear had been blocked by a ball of wax the size of a marble. The ringing stopped. I had no idea how deaf I had become until that moment when my hearing returned.

After he completed a thorough medical examination, the doctor said I was in great physical condition. "Wideman, you don't even have any cavities!"

My mental condition was yet to be determined.

A reporter wanted to interview me. The military approved, but the interview was tightly controlled. A full Air Force colonel stayed with me throughout. The journalist was a young man with long black hair, which caught me by surprise. I had seen the hippie look before, but it had not yet been embraced by many professionals when I left for Vietnam. This was my first indication that a lot could change in just half a dozen years.

The interview seemed to go well as far as I could tell. But at the end, the reporter gave the colonel an earnest look and said, "I'd love to ask him some real questions."

The colonel said, "I don't think he'll answer them."

"Try me," I said.

The colonel gave me a stern look, and the reporter didn't push it. What was the colonel afraid I might say? Had word of my reprimands already made it outside the 4th Allied POW Wing?

When I got home I saw the reporter's article in a St. Petersburg, Florida newspaper called *Evening Independent*. In it, he compared me to another returning prisoner, George T. Coker. I could see why the colonel had not wanted me to answer questions he wasn't warned about. In the first two paragraphs the reporter indirectly quoted us, saying of Coker, "He was ready to go back to a POW camp for American honor," and saying of me, "He'd never go back, and he seemed dubious about what America did in Vietnam." Although I was a little irritated because I had never spoken those words, he *was* right. If the colonel had given the reporter a chance to ask me how I felt about my service, I would have said just that: "I would never go back."

The reporter described me as a loner who said that during my time as a POW I made three or four "good friends and no enemies," and "did a lot of soul-searching." What was I thinking so deeply about? "Going back to school, getting out and accomplishing something," he quoted me. "Most of these people—the other prisoners—were doing what they wanted to do. It was part of their service."

By contrast, he described Coker as "a joiner," who praised the camaraderie of the flyers he had been imprisoned with. "I would be happy to do it again. I have no regret. We won a fabulous victory against communism. Those who say differently, simply do not understand what is going on."

I had been living both among men like Coker and men like me for several years, and I wondered which of us was the one who didn't understand what was going on? I had gone into the war believing that our military leadership was on a mission to defeat communism. But after living cheek-by-jowl with several military leaders in Vietnam's POW camps, I had developed a different picture, that some leaders do in war the same thing they do in peacetime: look for ways to increase or protect their power.

So the reporter painted an accurate, if incomplete, picture of the way I thought about the war. But he wasn't doing my reputation any favors by saying so.

I was at Clark Airbase only a couple of days before our group boarded another plane to return to the States. On this plane, a public affairs officer told me his version of the story about why the U.S. Government had bumped me from the special flight of twenty guys who went home out of shoot-down order. He started by repeating the story about Jim Bailey needing to go home to see his dying father. But then he said that, since Bailey was a Navy lieutenant, the one to be dropped had to be a Navy lieutenant. He said there were three Navy lieutenants in the group of twenty, so they drew straws. That's how I was picked, he said, short man out. I believed him.

As the years wore on, that reasoning seemed increasingly absurd. The initial premise for choosing which prisoners to send home, and when to send them, was shoot-down order. The first pilot shot down would be the first to go home. There were many prisoners in that list of twenty who were shot down *after* I was. Our government could have just picked the prisoner who was the last to be shot down and let him stay behind. Instead they picked me. As I came to realize the political importance of our small group of POWs, I found it less believable that a decision with international diplomatic implications would have been left to drawing straws or anything so random. I became convinced that someone had purposely selected me to remain behind. I'll never know who.

On the way home, our plane stopped at a naval base in Hawaii, where the base commander rolled out a long red carpet. When our individual names were called, we each walked the carpet to thunderous applause from a massive crowd. We were escorted to the operations building, where local families who lived on base met with us for about an hour. The people we met were very kind, but I was exhausted. After so many years of strictly limited contact with fellow humans, this parade of strangers was overwhelming for me. I might have preferred to be rolled up in that red carpet and tucked away in a dark room to sleep for a couple of days.

We soon boarded the plane again for the mainland, and the public affairs escort told me to smile and wave more the next time we deplaned in front of a crowd.

We made one more stop on the West Coast before a handful of us boarded a flight bound for points Northeast. The trip stretched into a sleepy blur of landings and takeoffs, but sometime after dark I landed at McGuire Air Force Base in New Jersey, the only prisoner left. A group of about three hundred people waited to greet me, mostly people from the base waving "Welcome Home" signs. I couldn't get over the feeling that the Navy dropped me off at night so I wouldn't get full news coverage because I disagreed with our leaders over the outcome of the war. Still, I stepped up to the waiting microphone and thanked everyone for coming to see me. I don't think I said much else. I was asleep on my feet, and my amplified voice floated to me from far away, maybe as far away as Vietnam.

At some point, I was whisked into the back seat of a black car, which drove off in an eight-car procession to the Philadelphia Naval Hospital.

I surveyed the line of cars and turned a quizzical look on the officer escorting me.

He smiled. "Those are for you."

I nodded, but didn't smile back. The endless journey home was taking it out of me.

At the hospital, where I was supposed to receive a more thorough inventory of my injuries, the press wasn't done with me. A gaggle of journalists waited in the lobby.

I stepped up to yet another mic: "I'm glad to be home, but I hope you'll understand I'm eager to see my wife."

Reporters called out questions, but I was already heading for the elevator. My wife was waiting somewhere above.

25

BREAKING BREAD

Inside the blessed silence of the elevator, my escort officer hit a button for one of the top floors, where he then led me to a large private suite. We walked into the living room, where just one person stood: my wife, Pat. After years of trying to remember her, imagine her, or reconstruct her from photos, it was strange to see her in three dimensions. I must have frozen for a second and stared, speechless.

The face I saw in my head did not exactly match that of the person before me. The twenty-year-old wife I'd left behind was now twenty-six and, if anything, she looked better than when I left. She did not rush forward and throw her arms around me, but stepped forward and gently hugged me. I wanted to feel like *This is it, this is coming home.* Instead, I felt uncertainty. Neither of us cried. I think she spoke, but I could not focus on what she was saying. It was all I could do to take in the reality that we were in the same room together.

The escort stepped outside to leave us alone, and I finally heard one sentence. "Bob, I'm glad you're home, but I'm sorry, I think our marriage is over."

I nodded, not agreeing, but just letting her know I understood the words.

Then she said that my family was waiting downstairs. "Your dad doesn't look well, so be prepared." I poked my head out into the hall and asked the escort officer to bring my family up.

A few minutes later, a group of footsteps approached from down the hall. I heard what could only be my mother's loud, wailing sobs, "Bobby! Bobby!" She sounded more like a mother in mourning than one welcoming home her long-lost son.

My parents, my brother Richard, and his wife Jan entered the room. Seeing my parents shocked me. My fifty-seven-year-old father looked at least a hundred. My mom had gained forty pounds. It was easier to look at Richard, who looked healthy and in good spirits. We all exchanged hugs, my mother clinging to me as if I might vanish again at any moment.

The Navy brought in a huge platter piled with food, which we barely touched. We sat down and tried to talk, but every subject seemed awkward: What had they been up to? What had I been up to? How could we answer such questions? My mom was the only one who kept up a constant stream of chatter. She said she was still working as a nurse, and that she had kept herself in good spirits by staying busy while I was gone. I wanted to believe that, but her face was etched with the seconds, minutes, and hours of worry she had endured every day of my imprisonment. My dad, who was divorced from my mom and seemed uncomfortable, maintained a stoic silence, staring at me with tears in his eyes.

Richard, Jan, and Pat chatted with an ease I didn't anticipate. Richard and my wife had barely known each other before I left, and this was my first time meeting his wife. From what I picked up, it seemed Pat had kept them all updated about me based on my letters from Vietnam.

After ten or fifteen minutes of this, I could not take any more. I excused myself to use the room's bathroom. It seemed my mom and dad had been destroyed the day my plane crashed. I stood in the bathroom for several minutes and cried. Then I splashed my face with cold water, toweled off, and returned, my surface composure restored.

I thought it would somehow get easier after that. It didn't. After an hour and a half, they all left. I walked into the room next door, the hospital room where I was expected to stay while physicians and psychiatrists decided whether I was healthy enough to go home, wherever that might be. I walked over to the window. It was open, the only thing between the ground and me a flimsy screen. The room was about forty floors up. I looked down and saw Pat walking to her car. She looked so alone. It would be easy to just jump out the window. Forty floors should do the trick.

It wasn't the thought of Pat's reaction that stopped me—it seemed to me she was done with me—but I thought about my parents and my brother, Richard. Now wouldn't that be the shits for them if I jumped out this window? I put the idea away.

In years to come I often wondered: why did the Navy put a POW in a top-floor room with only a screen between him and the ground?

A depressing sensation weighed on me, that it was not possible to come home after all because all that I had thought of as home was gone. I lay in my bed and tried to escape the feeling by falling asleep. But sleep eluded me.

Instead, I called for my new escort, Wade. That was one good thing the Navy did when the POWs returned home: arranged for each of us to have an escort. Wade was a perfect match for me. I told him I wanted to visit with my brother and his wife again, in private. "But please tell them I do not want to hear any opinions because I've had to listen to opinions for six years in prison."

Wade phoned Richard and Jan, invited them back, and gave them my edict.

The first thing my brother said to Wade when he walked into the lobby: "No opinions? How do you talk without expressing an opinion? You cannot talk without giving an opinion!" Later on, we had a good laugh over that.

Without my parents in the room, conversation came easier. Nothing important, just the ease of brothers who once knew each other by heart without saying a word. In that moment, I felt happy for my brother, whose life I had feared for all those years. He had made it out the other side of the war intact.

Wade brought back the rest of the food. We all admitted we were starving and ate it all.

As if seeing my brother happy was just the medicine I needed, I had my first decent night's sleep in years. What a simple pleasure it was to sleep in a room where I could turn out the lights whenever I wanted.

In the morning, while I shaved, the hospital's public affairs officer came into my room. He wanted to make sure I did not talk to any media without him present. I told him not to worry. I had no interest in speaking to the media.

In fact, I was reluctant to talk much with anyone. But that wasn't up to me. I was required to meet with the hospital psychiatrist for two hours every morning, Monday through Friday. Each morning, he asked me how yesterday had gone. So I kept telling him about yesterday, and then I would talk about my life as a prisoner. After two weeks of that, one morning I initiated the conversation. "So Doctor, how did your day go yesterday?" That was the last time we talked. I had a good laugh, all alone in my hospital room, about how I'd turned the tables on the psychiatrist. Apparently that was the day he decided I was mentally sound.

During my first two weeks back, I also went through a thorough intelligence debriefing. I liked the two people who debriefed me. They were thorough and asked smart questions. They recorded our conversations but turned off the tape recorder whenever I asked. They also let me read their print versions of all our conversations before they sent them to Washington.

During my comprehensive physical, I wasn't surprised to learn I had broken two fingers, which I had felt the moment I ejected from my plane six years earlier when my hand bashed into my helmet. A surgeon repaired the worse of the two. When he took off the cast and told me to move my finger, I felt blood rush from my face to my feet and almost passed out. I had gotten so used to that finger being useless, it never occurred to me I might use it again.

"Thanks, Doc," I said. "For six years, I've only been able to give everyone the finger with my right hand!"

The doctor told me I had a winged scapula, meaning one of my shoulder blades was out of whack because of damage to nerves that controlled the muscles around the bone. It probably happened during my ejection from the plane and was exacerbated when Vietnamese interrogators trussed my arms with ropes.

The doctor failed to find all my physical issues at the time. It would be decades before another doctor would tell me I had also suffered disc compression in my neck and lower back. That doctor described those disc compressions as "classic ejection injuries."

Commander Wilber was at the hospital as well. Most POWs were still following our order not to talk to him. I found out that the allegation against him was that he had cooperated with the Vietnamese by talking to foreign delegations in return for preferential treatment. I did not witness any of that, and after all that my superiors had put me through without cause I was more determined than ever to consider a man innocent until proven guilty. I had no problem talking to Wilber while we were both at the hospital.

Later, the Navy sent me to Washington to talk to Secretary of the Navy John Warner. Warner asked me if I had lived with Wilber during my time as a POW. I said no. I said that the other prisoners were mad at Wilber because they thought that he had received special treatment, but that the only preferential treatment anybody ever mentioned to me was that Wilber had accepted an extra banana. "I find it hard to believe that someone would cooperate with the Vietnamese just for an extra banana, sir." That was the truth.

I just wanted to get on with my life.

That sounded simple, but I soon discovered that getting on with even the simplest things in life was an undertaking that was beyond me. Wade took me to the Navy Exchange to buy clothes. It was a completely disorienting experience, and not just because I was a typical man who hated shopping. I had no clue what to buy. I stood in the middle of racks of clothes, totally out of it, unable to make sense of the huge selection after six years of having no choice about what to wear. I bought a pair of candy-cane-striped leisure pants, and as I walked out of the store with them it dawned on me that there was no way I was wearing those in public. I needed help.

"What's wrong?" Wade asked.

I remembered something Pat had said the night before: "I want you to know I still care about you, and I know you've been through a lot. So if you need anything, I'm happy to help."

I told Wade, "I need Pat to help me shop."

We went back to the hospital, where Wade called Pat for me and told her my dilemma. She agreed to go shopping with us that evening. This time we bypassed the Navy Exchange and went to the Cherry Hill Mall.

When we walked into the mall, it sank in that I was really, truly, back in the U.S.A. Why? I saw a black man in a blown-out afro, long trench coat, and wildly colorful clothes, and Pat tried to explain that the style was popularized by a recent movie called *Super Fly*, though she had no clear idea what the movie was about.

The *Super Fly* guy was just the first sign of how psychedelic my country had become. I saw bright colors everywhere: clothes, album covers, candles and incense. I bought some shirts and shoes that seemed a little wild to me, but which Pat assured me were actually on the conservative side. Seeing how much styles had changed made me feel like I had stepped into a time machine. There was a reason it felt as if I had been gone so long. Because I had.

When I picked out the shoes I wanted, I had no idea what to do next. I had to watch what other people did after they picked out their shoes. After I observed a couple of people go to the register and pay, I did the same. This time, though, the problem was not that the process had changed since I'd left, but rather that I had gotten out of the habit.

It wouldn't be the last time that would happen. For a couple of weeks, I had to watch other people do almost everything at least once before I could do it.

Pat went with Wade and me on a few more shopping trips. One night after Wade left us alone in my room, she gave me a long look and said, "You know, it just isn't fair that you spend six years in prison and when you finally come home I leave you. I think we should try to make it work."

We kissed. It wasn't romantic. It was awkward. It wasn't what I had imagined. But it was a start.

As a symbolic gesture of our fresh start, Pat and I decided to get remarried. We didn't waste time, but had a small ceremony in the hospital chapel.

At first we lived with Pat's parents in Cherry Hill, New Jersey, but I was determined to get on with the independent life I had been waiting for. So I pursued my long-delayed goal of getting my commercial pilot's license. I trained with the Air Force Reserves at Naval Air Station Willow Grove, Pennsylvania, and passed the written test.

License in hand, I lined up interviews with Delta, American, and Eastern Airlines. During my interview at Eastern, the chief pilot grabbed a stack of papers so thick he could barely hold it in one hand and thrust it at me like an accusation. "I've got three thousand applications here. They all have fifteen hundred-plus hours of jet-time. I can ask for PhDs and get them."

Delta expressed interest. But first, the company insisted I repeat advanced flight training with the Navy. For that, I went to Naval Air Station Kingsville, Texas.

I was in the program for a week when Eastern Airlines sent me a rejection letter. The letter said that due to my "ordeal," it would be better for me and for them if I re-applied in another year. That crushed me. It seemed cruel. When I came home, the Navy begged me to get back in the cockpit, but all Eastern saw was that I was a risk.

Meanwhile in Kingsville, a Marine major showed me the flight syllabus he had prepared for returning prisoners. It had a section requiring us to do carrier-landing qualifications. I told him I didn't need carrier qualifications because I wouldn't be required to do that sort of work again, not for the airlines. "Sorry, those are the rules for POWs," he said.

I took the issue up with the base commander, a Navy captain. He looked me up and down and said, "If you want to get off to a good start here, you need to get your hair cut and you need to wear a t-shirt." Between Eastern Airlines' rejection and this captain's petty requirements, I saw the writing on the wall. I was done.

The Navy offered to send me to the Naval Postgraduate School (NPS) in Monterey, California. To me, this represented the best opportunity to complete my education and get on with my life. So Pat and I moved to the coastal paradise of Monterey in the fall of 1973.

That's where I met fellow pilot and POW Tom Latendresse. The first time we talked, I told him about my effort to become a commercial pilot. Tom said, "This will really piss you off..." and told me how Delta Airlines begged him to apply for a job. He had been a later shoot-down and spent less time in prison, so Delta was willing to take a risk on him in return for the good publicity of hiring a prisoner of war. Even though I had sacrificed six years of my life, had a clean bill of physical and mental health, and the military seemed in a hurry to send me back into the sky, the airlines were less inclined to take a chance on someone who had spent so long languishing in a cell.

At NPS, I pursued a Bachelor of Arts in International Relations. I was itching to learn anything to help me understand why I had spent six years in a communist prison. I learned about world politics and came to see that the reason we lost Vietnam was pretty simple: America's leaders had the hubris to assume that Vietnam was just some simple little backwater country we could rush in and defeat without trying to understand its history, culture, or people. America's leaders had gotten so used to winning that they assumed we would always win. We have paid a heavy price for that arrogance.

Shortly after the war ended, Congress discussed crediting Vietnam POWs with two years for each year we spent in prison, to go toward our retirement. Two POWs told me that General Robinson Risner and Admiral James Bond Stockdale appeared before Congress and testified that we did not need extra consideration because we had served for patriotic reasons. I could not believe they had the arrogance to speak for all of us. Both of those guys already had great pensions. Why didn't they want other POWs to get a good deal too?

James Stockdale was awarded the Congressional Medal of Honor because he cut his wrists while in prison. I remember being told through the walls of one of the camps that Stockdale had cut his wrists. But his award indicated that his suicide attempt on September 4, 1969 resulted in better treatment for the rest of us prisoners, saving some of our lives. I have never experienced, witnessed, or heard any evidence of that. Except for some rough punishment in May and June of 1969, after the escape attempt, our mistreatment had stopped in September of 1968, after Johnson stopped bombing North Vietnam in an effort to help get Hubert Humphrey elected president. In other words, the bad treatment had stopped long before Stockdale attempted suicide. No one I've asked has ever been able to name one prisoner whose life was saved as a result of Stockdale's attempt to end his own life.

I cannot speak for the POWs of South Vietnam. Even when it comes to North Vietnam, I only speak for myself. But if the other POWs in North Vietnam experienced what I did, then the war was harder on our families back home than it was on us. We knew we were getting our two slops and a flop every day, while our loved ones had no clue what was happening to us. They only had the words "prisoner" and "war," and their imaginations. The human imagination is a gift, but for the families of POWs it was cruel and unusual punishment.

Shortly after I came home, my mom and I went out to dinner in Ricky River, Ohio. We both ordered steaks. In fact, when I first came home I ate steak three or four nights a week for six months, just because I could. The moment the waiter set down our plates, I started to carve and eat my filet with gusto. I was almost halfway through when I noticed that my mom was just sitting in her chair, staring at her steak. She hadn't touched it.

"Mom, why aren't you eating your steak? It's delicious."

She looked up at me with hurt in her eyes. "You know, Bobby, the whole time you were gone I could not eat a steak like this because I didn't know if you were getting enough to eat."

I put my fork down, and Mom reached out to touch my arm. For a long while we both sat silently, just looking at each other, not eating a bite.

AFTERWORD

A Different POW Experience

The POWs who landed in Hanoi's prison camps can thank God their treatment was as good as it was. I know some never saw it that way. Only seven prisoners died in Hanoi: two stopped eating; one died from a combination of ejection wounds, exposure, and the Vietnamese rope treatment; one died during an escape attempt; and one succumbed to typhoid. I'm not sure what happened to the other two.

In America's Civil War, thirteen thousand Union prisoners died at the Confederacy's infamous Camp Sumter near Andersonville, Georgia. In World War Two, the Japanese chopped off two American heads for every mile of the sixty-five-mile Bataan Death March. Of the more than twenty-seven thousand American POWs in Japan, between 27 and 40 percent died in captivity. In that same war, Germany admitted that three million Russians died in German prison camps. In turn, the Russians captured ninety-five thousand Germans at Stalingrad and only four thousand returned home.

With the exception of some of America's prisoners in World War Two, it may be that never in the history of warfare have POWs been treated so well as we were in North Vietnam. Prisoners held by the Viet Cong in South Vietnam were another story; I won't speak to that because I wasn't there.

Although I suffered painful physical punishment, which some call torture, I've always had a hard time calling what the North Vietnamese did to me torture. It was a bad experience, but it could have been much worse.

Although we successfully established communication at each prison camp, it was not perfect or consistent. Many POWs later talked about how we were always able to communicate despite the North Vietnamese Army's efforts to stop us, presumably because of the "great leadership" we had. On the contrary. The NVA leadership proved they could shut down our communications whenever they wanted, which they did after the escape attempt. Some key personnel did not communicate for two months.

It was clear to me that many Naval Academy graduates and senior officers did whatever it took to please their bosses. Such sycophants taught me one of the most important lessons I learned from my Vietnam experience: there will always be people who pursue power by ingratiating themselves to those in power without pausing to assess the goals of those leaders. I came to understand this as a POW, but I have witnessed it in all institutions since: corporations, bureaucracies, schools, churches, you name it.

My sense is that most pilots had huge egos—me included—which probably drove us to become fighter pilots in the first place. The most hardline of the POWs had the most problems in prison. The North Vietnamese forced them to make the most confessions and visit the most delegations to feed the Vietnamese propaganda machine.

It's well documented that many American political and military leaders knew we were fighting an unwinnable war but said nothing because they feared jeopardizing their careers. Those same leaders demeaned and discredited the courageous Americans who publicly opposed the Vietnam War, especially big names like Jane Fonda. When Fonda came to visit us in 1972, we were being treated well, just like she said we were. We went outside several hours a day, ate three meals a day, and received regular letters and packages from home. The barrage of war protests put pressure on the government to end the war. But for them, we would still be over there.

When we came home, POWs who supported the war were encouraged to speak out while those who did not were not encouraged to speak out. That policy continues today, and is one reason we have an inflated view of the importance of funding America's military might. We primarily receive the viewpoint of those invested in maintaining power.

After the war, I talked to an Army colonel in Tampa, Florida who helped plan the Son Tay Raid. He told me that the American military knew the camp was empty thirty days before the raid, but our leadership weighed the costs and benefits of going through with it anyway, and the benefits won. They knew they would recover no prisoners. Such was the American need to keep its own propaganda machine running.

A Wartime Nation

Our armed services have not won a conflict since World War Two, yet we keep waging war as if it were the national pastime. One reason this happens is because so many of our military leaders want to perpetuate their power.

Little has changed in the military since we lost in Vietnam. We continue to pursue costly wars that yield questionable results. The invasions of Iraq and Afghanistan, like Vietnam, were monumental blunders motivated by American hubris. Once again, we have preyed on countries that we view as weaker than ours and have tried to impose our will on them, only to discover that the will of other cultures to chart their own course is stronger than we anticipated.

In Vietnam, we supported a Catholic puppet regime even though 95 percent of the Vietnamese population was Buddhist. What made us think they would welcome us as liberators? Once again, we have installed puppet regimes in Iraq and Afghanistan, only to see fringe groups like ISIS take advantage of the power flux to inflame those disenfranchised by our interference. The local populations of those countries now hate us just as the Vietnamese did.

When I first returned from Vietnam, plenty in the military refused to let go of the belief we had won, despite the facts. They said things like: We stopped them...Our bombing campaign brought them to the table...It was a victory for America. Many bureaucrats and politicians do the same today, ignoring facts so they can cling to claims of success in Afghanistan and Iraq.

What's more, all of these wars have contributed to national inflation and debt, as well as international economic instability. President Johnson tried to initiate The Great Society and fight the Vietnam War at the same time. He had enough money to pursue one agenda, not both. President Nixon once admitted that one reason the Vietnam conflict dragged on was because he didn't want to be the first American president to lose a war. The reason we got out of that war wasn't because the U.S. was ready to admit defeat but because we couldn't afford it anymore.

President Carter inherited the inflation caused by Vietnam. Every economic crisis since has been aftermath. President Reagan said he would increase employment and kill inflation, even though economists said we couldn't have it both ways. A lot of people were impressed because he did it. How? He put everything on a credit card. That's when our debt started to skyrocket.

President Clinton made a dent in that debt, but President George W. Bush went to war and ran it back up again, from five trillion to ten trillion. Like the leaders who ignored the facts on

Vietnam, Bush ignored the facts on Iraq. Iraq did not perpetrate the 9/11 attacks. There was no Al Qaeda in Iraq until after Bush invaded. There was no evidence of weapons of mass destruction in Iraq. Bush had an agenda to take on Saddam Hussein, so he did, despite the facts.

Why would any thinking president take us into Afghanistan? The British went there and got their butts kicked. The Russians went there and got their butts kicked. Why did Bush ignore history? Someone once said, "Afghanistan is a place where great powers go to get humiliated."

Some generals warned Bush he couldn't win in Iraq with his limited troops, so Bush sought other generals who toed the party line and put them in charge. How else could General Casey have become a four-star general with no combat experience?

Meanwhile, the housing bubble burst in 2008 and our debt went up again. Today it has surpassed eighteen trillion dollars. The wars in Afghanistan and Iraq contributed to this debt.

The United States spends more on military defense than the top seven to nine nations combined, depending on which source you consider. The problem is not Persian warships in Chesapeake Bay. The problem is American warships in the Persian Gulf. We just keep sticking our nose into other people's business.

I've learned that almost every modern war is about lining pockets. I'm all for capitalism, but I know who stands to benefit if we convert the world to capitalism: big business. We had to kill the commies because they were going to interfere with America making money. Now we kill Muslims for the same reason. Nobody talks about it because it's not politically correct to ask people to die for money. Instead, leaders put a spiritual spin on it and make it a righteous cause.

In the military, the desire for money translates into the desire for power. That thirst trickles down through the ranks. I saw this firsthand in the POW camps.

It's popular to talk about these wars as fights for freedom or democracy, or as battles against political tyranny or religious fanaticism. It really isn't about religion or democracy. It's about rich versus poor. Of course, if we're talking about the soldiers on the front line, then it's simply poor versus poor. Those are the people fighting each other.

For Vietnam, we had a draft, but if a draftee's family had money he could get around that. We tried to stop that problem with the volunteer army. But who volunteers? The poor, who have few opportunities besides what the military promises. Different path, same result. The poor are the people we fight, and the poor are the people who fight for us.

Torture, American Style

We're all aware of the Bush administration's approval of the CIA torturing suspected terrorists at black sites around the globe. According to the Associated Press, the Congressional Record, Human Rights Watch, and the U.S. military's investigative documents, as of 2006, at least 108 POWs from our wars with Iraq and Afghanistan died in American custody. At least thirty-four of those deaths were either suspected or confirmed homicides. That's more than four times the number of the American POWs who died in Hanoi.

Bush's attorneys lined up experts who said that the CIA's "enhanced interrogation techniques" were not torture. Those techniques continued under President Barrack Obama. In 2015, the Senate Intelligence Committee commissioned a report on the CIA's interrogations and concluded that much of what had been approved does indeed constitute torture.

Here are just a few instances of torture that the report identified: One prisoner froze to death after being left to sleep without pants on a cold concrete floor. Another was forced to stand in a "stress position" on broken bones. Others were placed in isolation or were sleep-deprived until they suffered symptoms of psychosis such as hallucinations, paranoia, and self-mutilation. Some prisoners were forced to go through rectal hydration or rectal feeding, in which water or food was forced into the anus, which can leave the kind of damage associated with sexual assault. And of course, we've all heard the debates over waterboarding.

I agree with Senator John McCain's assessment of the report on two counts: 1) those techniques are torture, and 2) those techniques do not work. I have a problem with our country torturing war prisoners, both because it is morally wrong and because it creates more enemies for America. We call what our enemies do to their prisoners torture while asserting that we're a kind, just people who don't do that sort of thing. I find it offensive that some POWs have supported the torture of prisoners in the Iraq and Afghanistan Wars after whining about their own treatment by the North Vietnamese.

In any case, there's no need to go so far. The Vietnamese got all the information they needed by bringing people to a certain point of pain and holding them there. Beyond that point, people will say or do anything. That's when information becomes unreliable. Our country has inflicted prisoners with torments well beyond anything I suffered in Vietnam.

In my opinion, the people who order the sort of torture described in that Senate Intelligence Committee report are war criminals, Bush and Obama among them. I consider the subordinates who carried out those orders guilty too. I believe we must each take responsibility for the morality of our actions. We need to try all of them for war crimes.

Divorce Epidemic

Not two years after the North Vietnam POWs returned, the divorce rate among our ranks soared to 85 percent. This high number was likely a result, at least in part, of post-traumatic stress and the long separation of husbands and wives.

Pat was 19 and I was 22 when we married. We were just too young. A few weeks after I returned home and we took our vows again, we visited another couple. The wife pulled me aside to say, "Don't be so critical of Pat." She was right. I was very impatient with my wife.

Excessive arguing is a classic symptom of Post-Traumatic Stress Disorder, which I didn't know much about at the time, but which soon became a household word surrounding the subject of Vietnam War veterans. Pat and I argued so much that our seventy-pound Doberman Pinscher hid behind the sofa. More importantly, we had two sons: Eric, born in 1974, and Derek, born in 1976. What did those arguments do to the minds of a two-year-old and a four-year-old?

In 1976, we left our home in the beachside town of Monterey, California for Meridian, Mississippi, where I became a Navy comptroller. The arguments escalated. Neither Pat nor I wanted to hurt our boys. We soon separated, and in January of 1978 we divorced.

On My Knees

From long before Vietnam until long after, I didn't believe in God. I considered God an imaginary crutch for people too weak to handle their problems. I realize now that toughing out imprisonment without any spiritual support inflated my ego.

After I separated from my wife, I knew I needed to talk to somebody. I thought I had two choices: a minister or a shrink. I remembered that Thomas Eagleton underwent psychiatric treatment before he ran for Vice President as McGovern's running mate in 1972. The media got wind of that and crucified him as if it meant he were crazy. That stigma convinced me to avoid psychiatrists and psychologists. I didn't want any chance of this difficult period coming back to bite me. I talked to the base chaplain.

I told the chaplain that I had long ago given up on the idea of God. He recommended I try a few different Protestant churches and advised me to read *The Gospel of John*, a gentle introduction to the Lord after being away from Him for some twenty years.

I read *John*, but he made no sense to me, not then. On the advice of a friend, I added the writings of Carlos Castaneda to my reading list. Castaneda made me aware of how much the ego is in charge of our lives, via the constant refrain: "I want." I not only studied the Bible, but also read about Buddhism, Judaism, and philosophy. I saw that everything came down to ego. I noticed the word "I" rarely appeared in the Jewish teachings of the Pirkei Avot, or Ethics of the Fathers, one of the texts in the compilation of rabbinical wisdom called the Mishnah. This brought to my attention that the Jewish people I knew did not use the word "I" very much. I sought to reduce the use of the word "I," and found that my boss and others listened. It was a transformation.

I ultimately landed in a Southern Baptist Church. I knew I could never toe the entire party line of any organization, but the Southern Baptists and I were on the same page about focusing less on "self" and more on "we"—on community.

One day in 1991, I had an epiphany about *The Gospel of John*: I could forgive other people's sins but I did not have the power to forgive my own. I realized only Jesus Christ had that power. The day I understood that, I dropped to my knees and forgave everyone I could think of who I felt had ever wronged me. That had a huge impact on me.

Among the people I had the biggest beef with were a few of my fellow prisoners from Vietnam, particularly our leadership. In my prayers, I forgave all of them, even the ones who wrongly accused me or humiliated me. Forgiving the North Vietnamese was never an issue because I always thought they could have treated us so much worse.

My trials as a POW did not bring me to God. Getting divorced did. It surprises some people when I tell them getting divorced was more stressful for me than being a POW.

A Bad Reputation

I first attempted to write a book about my war experience in the mid-1970s, but I fictionalized it as a novel. I sent a manuscript to the Naval Investigative Service, because the Navy required me to get their approval. It turned out that the mere act of seeking approval was enough to get me in trouble.

Several months later, the Naval Investigative Service sent the book back, not to me but to the superintendent at the Naval Postgraduate School. The cover letter called my book inaccurate, immature, and demeaning to fellow POWs who deserved to be lifted up. I had done nothing worse than paint all of us POWs as people instead of saints. What upset me more was that my pages came back in complete disarray. I had accorded the Navy the respect of requesting approval, and in return I had received a slap in the face.

I called the Naval Investigative Service to ask what the letter meant. The person I spoke to informed me that if I published the manuscript it would only serve to publicly discredit me.

"According to whom?" I asked.

He said, "We didn't know what to do with your manuscript, so we sent it to your roommates and to Stockdale."

"You did what?!" I had sent my manuscript to the Navy in confidence, and someone had published it to other people without my consent.

A few years later, I was passed over for promotion to commander. When I inquired to learn why, my detailer in Washington said, "You need to call Stockdale. He told the board you had a bad reputation." Stockdale was the head of the promotion board.

"A bad reputation for what?" I asked.

He said Stockdale gave no specifics. I was stunned at the abuse of power this implied: that the board listened to him despite not having evidence against me. This was all the more suspicious because shortly after I had returned from Vietnam, Stockdale had submitted a fitness report in which he recommended me for promotion. His change of heart came *after* I submitted my manuscript to the Navy.

Years later, in 1996, I called Stockdale and asked if there was something I should know. He told me that the promotion board was just a "paper push" and that he said nothing detrimental about me. He claimed he had never seen my manuscript.

Stockdale and other POWs wrote books about their prison experiences, but their books painted the military in a more glowing light. The Navy never sent their manuscripts to me for my response even though someone sent my manuscript to Stockdale and my roommates for their comments and approval. The double standards of my POW days continued.

About three or four years ago, when I was in Pensacola, Florida, I told the head of the Robert E. Mitchell Center for Prisoner of War Studies, "I got passed over for commander because Stockdale told the board I had a bad reputation."

He looked me in the eye and said, "I can promise you, you don't have a bad reputation among the five or six hundred prisoners from North Vietnam."

Years later, I talked to a doctor from the Mitchell Center who was a friend of Stockdale's and who made it clear he truly liked the man, and he told me, "Stockdale would do something like that." To this day, many senior leaders are big on protecting their turf and their reputations and not averse to tearing other people's reputations apart to achieve that.

When I talked to the base chaplain about my divorce, we also talked about the war, and he called me a conscientious objector. I had never thought of myself that way, but he had a point: I didn't believe in the war anymore. I had heard our leaders distort facts to make themselves look good. I never publicly protested—it was too late for that—but I got demerits for not agreeing that we achieved "a fabulous victory against communism."

In the end, Stockdale's pursuit of power took him all the way to vice admiral. Meanwhile, he claimed I had a bad reputation. All he had to do was say the words, and because he was one of the highest-ranking officers in the military, people on the promotion board believed him.

A Changed American Dream

The Navy sent me to the Naval Postgraduate School to get an undergraduate degree in International Relations. After that, my superiors urged me to get back in a cockpit, saying that was the route to make command. The military had not been my dream, so I pursued a master's degree in finance. With that, the Navy wanted to send me to Washington as an auditor. I didn't want that. Instead, I went to the Naval Air Station in Meridian, Mississippi to become a comptroller for seven years.

I retired from the U.S. Navy in 1983 and went to Florida to work as a stockbroker. I never got over the feeling that prison had cost me years of time and opportunity, so I went on to earn a law degree at the University of Florida. I became a prosecutor in 1991, and a few years later went into private practice. However, the sedentary nature of that career sent my blood pressure up. Then, in 1996, the Navy made me an offer I couldn't refuse: I moved back to Mississippi to become a flight simulator instructor. Flying for the Navy had landed me in prison and stolen years of my life, but training other pilots turned out to be one of the best jobs I ever had. I worked as an instructor until I retired in 2012. I was 68.

Perhaps one attraction of training pilots was that I never completely got over my frustration at not becoming an airline pilot, the dream I had held onto during my six years as a prisoner of war. Being rejected by Eastern Airlines was more devastating than anything the North Vietnamese could have done to me.

The Test of a Man

When I consider how capable we all are of perverting the truth, and when I remind myself that I was a voluntary participant in the Vietnam debacle, I can only ask: what does it take to be a man? I submit that a real man is not a sycophant, but is someone who pursues the truth in service to his values. It's easy to support the status quo when self-interest is at stake. It takes character to stand up for the truth when it's not in your self-interest—such as opposing war in the face of threats to destroy your reputation.

It also takes character to apologize when we're wrong, which is something the U.S. has yet to do for Vietnam. We invaded their country and killed more than two million Vietnamese because a majority of them did not want us to tell them what kind of government to support.

Best I can figure, humans point their fingers at others when they need a scapegoat. Usually they point at someone with less power because that's easiest, to draw attention away from their own shortcomings. Once the finger pointing starts, honesty is the first casualty. When honesty goes, everything goes. To me, this was not only the dynamic between the leadership and the subordinates among the POWs, but also the dynamic between the U.S. and Vietnam. We saw them as less powerful, so we thought they were an easy target. We were wrong.

Despite the pitfalls of ego I saw many military leaders display in Vietnam, I find it important to remember the exceptions, men who provided a standard for honest reflection on right and wrong action, and who were not afraid to engage in criticism—of authority or of themselves—when honesty called for it. I have tried to introduce some of those men to you in these pages. Perhaps some in our POW leadership felt justified in attacking men who believed in following conscience first and orders second, but what really made such men targets was that they had no rank and no power and seemed easy to suppress.

Studying the teachings of Jesus has taught me the importance of placing truth above pride. Wars go on, but I have found peace in this: "Love God with all your heart, mind, and soul, and love your neighbor as yourself." This has been and continues to be my journey, and it is one reason I've chosen to share with you the sometimes-painful story of my experience as a prisoner in Vietnam. I hope my story helps open your heart to the challenge of getting to know yourself, your fellow humans, and the people who share your world.

Soldiering On

I still think about war and imprisonment, their causes and consequences. It's part of being an informed person, and my experiences have helped to make me an informed person. But my life has also been filled with blessings: two children, six grandchildren, true friends, education and the opportunity to pass it on, fruitful labors, the freedom and means to travel, good health, and a relationship with the Lord.

Sometimes it's painful to remember my six years of lost freedom, being isolated from loved ones while at the same time discovering the truth behind Jean Paul Sartre's words: "Hell is other people." Most of the time, those memories remind me to be grateful for my life now.

When I came home, I was so happy to be free that I let go of the suffering I had endured. As I grow older, the past creeps up on me. Most memories fade while others intensify over time. Most of the time I can accept this. Other times I have to pray. With post-traumatic stress, that's just the way it goes. After the war is over, after we become civilians again, still we soldier on.

ACKNOWLEDGMENTS

This book would not exist without my family, friends, and many fine fellow POWs. Thanks to all of you for seeing value in my story and in me. A special thank you to Richard for being the great brother he always has been and for believing in me. Many thanks to my daughter-in-law, JD, for her support and encouragement throughout this long journey. Thanks to Celeste Currier for her friendship, encouragement, and moral support. I'm grateful to the late Bradwell Scott for being a true friend, for believing in me, and for listening to my story.

My mom and dad have passed on, but I will always be grateful to them for their unwavering belief in me. Their faith helped me make it through the hard times, and beyond.

Thank you to those fellow Vietnam POWs who made my time in prison more bearable, those who supported me in the years that followed, and those who encouraged me to write this book. Thanks to my friend Howie Hill who generously helped me recall details of his story. Thanks to my ex-wife, Pat Stuckert, who waited for me and did her best to be there for me during a difficult time.

I also want to thank a man who saved my life: PR1 B.L. Simpson expertly maintained the ejection seats and perfectly packed the parachutes of all five of the pilots in my squadron who were forced to eject during my last cruise. He batted a thousand: all five of us survived thanks to him.

CPSIA information can be obtained
at www.ICGtesting.com
Printed in the USA
LVOW12s2014040716
495088LV00001B/40/P